AF564903

Biotech Laboratories, Experiments, Equipments & Institutions

Biotech Laboratories, Experiments, Equipments & Institutions

P. R. Gilbert

ANMOL PUBLICATIONS PVT. LTD.
NEW DELHI - 110 002 (INDIA)

ANMOL PUBLICATIONS PVT. LTD.
H.O.: 4374/4B, Ansari Road, Darya Ganj
New Delhi-110 002 (India)
Ph.: 23278000, 23261597
B.O.: 1015, Ist Main Road, BSK IIIrd Stage
IIIrd Phase, IIIrd Block
Bangalore - 560 085 (Karnataka)
Ph: 080-41723429, Tel/Fax: 080-2672 3604
Visit us at: www.anmolpublications.com

Biotech Laboratories, Experiments, Equipments & Institutions

First Published, 2008

ISBN 978-81-261-3279-9

PRINTED IN INDIA

Printed at Mehra Offset Press, Delhi.

Contents

Preface

This book entitled "Biotech Laboratories, Experiments, Equipments and Institutions" is designed to acquaint readers with the main biotech laboratories, experiments, equipments and institutions. Biotechnology involves mainly the industrial application of living organisms and/or biological techniques developed through basic research. Biotechnology products include pharmaceutical compounds and research materials. Primarily it is the scientific manipulation of living organisms, especially at the molecular genetic level, to produce useful products. Gene splicing and use of recombinant DNA (rDNA) are major techniques used. In broader sense it is the technology for working with biological systems. Includes genetic engineering, human and veterinary medicine, crop and animal breeding, diagnostics, pharmaceuticals, forensics, etc. In narrow sense it means genetic engineering which could be understood as the manipulation of an organism's genetic endowment by introducing or eliminating specific genes through modern molecular biology techniques.

A broad definition of genetic engineering also includes selective breeding and other means of artificial selection. Molecular modeling is a technique for the investigation of molecular structures and properties using computational chemistry and graphical visualization techniques in order to provide a plausible three-dimensional representation under a given set of circumstances.Major biotech lab equipments include: Laser Capture Microdissection (LCM) and Laser Cutting Systems; RNA Isolation & Amplification Kits

(Formalin-Fixed Paraffin-Embedded Tissues); RNA Isolation & RT Kits (Formalin-Fixed Paraffin-Embedded Tissues); RiboAmp RNA Amplification Kits (Frozen Tissues); DNA and RNA Isolation Kits (Frozen Tissues); and Staining Kits (Frozen Tissues). Main products and processes used in biotech research include the Arcturus Laser Capture Microdissection; Cellular Imaging; Electrophysiology ; Fluorometric Imaging Plate Readers; Liquid Handling; MetaMorph; Microarray Analysis; Microplate Readers; Reagents; Software and Validation. Safety management and operation of processes in biotechnology laboratories also invite certain safety regulations.

This book consists of—Biotechnology Laboratories: An Overview; Biotechnology Experiments: Upcoming Areas; Biotechnology Research Related Equipments; and Major Biotech Institutions and Research Centres Around the World etc. besides a large body of list of acronyms, glossary of relevant terms, extensive bibliography, list of websites and links, for further referencing and research.

—Editor

List of Acronyms

AAAS	—	American Association for the Advancement of Science
AACC	—	American Association for Clinical Chemistry
ABIC	—	Agricultural Biotechnology Intl. Conf.
ABRF	—	Association of Biomolecular Resource Facilities
ACMG	—	American College of Medical Genet.
ACS	—	American Chemical Society
ADA	—	Americans with Disabilities Act
AES	—	American Electrophoresis Society
AFMR	—	American Federation of Medical Research
AFIP/ARP	—	Armed Forces Inst. of Pathol./Am. Registry of Pathol.
AGSG	—	Alliance of Genetic Support Groups (now Genetic Alliance)
AGT	—	Association of Genetic Technologists
AHA	—	American Heart Association
AIBS	—	Am. Inst. of Biol. Soc.
AMIA	—	American Medical Informatics Association
ANGIS	—	Australian National Genomic Info. Service

ANL — Argonne National Laboratory
API — Application Programming Interface
ARP — American Registry of Pathology
ASB — American Society for Biotechnology
ASBMB — American Society for Biochemistry & Molecular Biology
ASCB — American Society for Cell Biology
ASCI — American Society for Clinical Investigation
ASHG — American Society for Human Genetics
ASIP — American Society for Investigative Pathologists
ASIS — American Society for Information Science
ASLME — American Society of Law, Medicine, and Ethics
ASM — American Society for Microbiology
ASPET — American Society for Pharmacology and Experimental Therapeutics
ATCC — American Type Culture Collection
ATP — Advanced Technology Program
AVS — American Vacuum Society
AWCH — Adelaide Women and Children's Hospital
BAC — Bacterial Artificial Chromosome
BACR — British Association for Cancer Research
BCM — Baylor College of Medicine
BDG — Batten's disease gene
BER — Biological and Environmental Research

BIO — Biotechnology Industry Organization
BNL — Brookhaven National Laboratory
BSCS — Biological Sciences Curriculum Study
BTCI — BioPharmaceutical Technology Center Institute
BTP — Biotechnology Training Programs
CAE — capillary array electrophoresis
CATCMB — Center for Advanced Training in Cell and Molecular Biology
CCM — Chromosome Coordinating Meeting
CDC — Centers for Disease Control
cDNA — Complementary DNA
CE — Capillary Electrophoresis
CEPH — Centre d'Etude du Polymorphisme Humain
CF — Cystic Fibrosis
CFF — Cystic Fibrosis Foundation
CHH — Cartilage-hair Hypoplasia
CHI — Cambridge Healthtech Inst.
CHOP — Children's Hospital of Philadelphia
CIMB — Center for International Meeting on Biology
CIRB — Colorado Institute for Research in Biotechnology
CLMA — Clinical Laboratory Management Association
cM — Centimorgan
CMT — Charcot-Marie-Tooth
CONTIG — Consortium of Teachers in Genetics

CORN	—	Council of Regional Networks for Genetic Services
CRADA	—	Cooperative Research and Development Agreement
CSHL	—	Cold Spring Harbor Laboratory
CTG	—	Trinucleotide Repeat
DHHS	—	Dept. of Health and Human Services
DIMACS	—	Center for Discrete Math. & Theoretical Comp. Sci.
DM	—	Myotonic Dystrophy
DMD	—	Duchenne Muscular Dystrophy
DNA	—	Deoxyribonucleic acid
DOE	—	Department of Energy
ECB	—	European Congress on Biotechnology
EDS	—	Electronic Data Submission
EEOC	—	Equal Employment Opportunity Commission
EL.B.A.	—	ELectronics and Biotechnology Advanced
EMBL	—	European Mol. Biol. Lab.
EMBO	—	European Molecular Biology Organisation
EMG	—	Encyclopedia of the Mouse Genome
EMS	—	Environmental Mutagen Society
EORTC	—	European Organization for Research and Treatment of Cancer
ERDA	—	Energy Research and Development Administration
ERI	—	Eleanor Roosevelt Institute
ES	—	Embryonic Stem

ESF	—	European Science Foundation
ESHG	—	European Society of Human Genetics
ESI	—	Electrospray Ionization
EST	—	Expressed Sequence Tag
EUCIB	—	European Collaborative Interspecific Backcross
EURESCO	—	European Research Conferences
FAP	—	Familial Adenomatous Polyposis
FASEB	—	Federation of American Societies for Experimental Biology
FCM	—	Flow Cytometry
FEBS	—	Fed. of Eur. Biochem. Soc.
FMF	—	Familial Mediterranean Fever
FRAXA	—	Fragile X locus
FSHD	—	Facioscapulohumeral Muscular Dystrophy
FTICR	—	Fourier Transform Ioncyclotron Resonance
FVEA	—	Fundacion Valenciana de Estudios Avanzados
GAS	—	Genome Automation System
GBASE	—	Genome Database of the Mouse
GBR	—	Global Business Research
GDB	—	Genome Database
GDB/OMIM	—	Genome Database/Online Mendelian Inheritance in Man
GESTEC	—	Genome Science and Technology Center
GIST	—	Genome Informatics System of Transputers

GLaRGG	—	Great Lakes Regional Genetics Group
GMCRF	—	General Motors Cancer Res. Foundation
GMD	—	Genomic Map Design
GPI	—	Genetics and Public Issues Program
GRAIL	—	Gene Recognition and Analysis Internet Link
GRC	—	Gordon Res. Conf.
GSA	—	Genetics Societies of America
GSDB	—	Genome Sequence Data Base
HAEC	—	Human Artificial Episomal Chromosome
HELSRD	—	Health Effects and Life Science Research Division
HERAC	—	Health and Environmental Research Advisory Committee
HGCC	—	Human Genome Coordinating Committee
HGM	—	Human Genome Meeting
HGMIS	—	Human Genome Management Information System
HGP	—	Human Genome Project
HHMI	—	Howard Hughes Medical Institute
HICSS	—	Hawaii Intl. Conf. on Systems Sci.
HLA	—	Human Leukocyte Antigen
HMDP	—	Homology Database
HNPCC	—	Hereditary Nonpolyposis Colorectal Cancer
HSA	—	Human Serum Albumin
HUGO	—	Human Genome Organisation

IARC	—	International Agency for Research on Cancer
IBC	—	International Business Communications
IBEX	—	International Biotechnology EXpo
IBF	—	International Business Forum
IBI	—	Institute for Biotechnology Information
ICES	—	International Council of Electrophoresis Societies
ICGEB	—	International Centre for Genetic Engineering and Biotechnology
ICHG	—	International Congress of Human Genetics
ICPEMC	—	International Commission on Protection Against Environmental Mutagens and Carcinogens
ICRF	—	Imperial Cancer Research Fund
IEEE	—	Institute of Electrical and Electronics Engineers
IG	—	IntelliGenetics
IGES	—	International Genetics Epidemiology Societies
IJCAI	—	International Joint Conference on Artificial Intelligence
IMA	—	Institute for Mathematics and its Applications
IMACS	—	International Association for Mathematics and Computers In Simulation
IMAGE	—	Integrated Molecular Analysis of Gene Expression

IMGS — Intl. Mammalian Genome Society

INRIA — French Natl. Inst. for Research in Computer Science and Control

IOR — Institute of Religion

ISAG — Intl. Society for Animal Genetics

ISMB — Intelligent Systems for Molecular Biology

ISONG — Intl. Soc. of Nurses in Genet.

ISQL — Interactive Standard Query Language

ISTR — Institute for Science Training and Research

IUBMB — Intl. Union of Biochemistry and Molecular Biology

IU — Indiana University

IVD — *in vitro* Diagnostic

JGI — Joint Genome Institute

LANL — Los Alamos National Laboratory

LBNL — Lawrence Berkeley National Laboratory

LLNL — Lawrence Livermore National Laboratory

LTI — Life Technologies, Inc.

MALDI — matrix-assisted laser desorption ionization

MBC — Massachusetts Biotechnology Council

MBL — Marine Biological Laboratory

MCD — Mouse Cytogenetic Database

MDA — Muscular Dystrophy Assoc.

MEMS — Microelectromechanical Systems

MGC — Mouse Genome Conference
MGD — Mouse Genome Database
MGI — Microbial Genome Initiative
MHC — Major Histocompatibility Complex
MIMBD — Meet. on the Interconnection of Mol. Biol. Databases
MIT — Massachusetts Institute of Technology
MMRF — Marshfield Medical Research Foundation
MNBWS — Miami Nature Biotechnology Winter Symposium
MOD — March of Dimes
MOU — Memorandum of Understanding
MRC — Medical Research Council
MS — Mass Spectrometry
NACHGR — Natl. Advisory Council for Human Genome Res.
NAPBC — Natl. Action Plan on Breast Cancer
NAS — Natl. Academy of Sciences
NASA — Natl. Aeronautics and Space Administration
NBAC — Natl. Bioethics Advisory Commission
NCGR — National Center for Genome Resources
NCI — Natl. Cancer Institute
NCSA — National Center for Supercomputing Applications
NCSL — National Conference of State Legislatures

NCSU	—	North Carolina State University
NEH	—	National Endowment for the Humanities
NFCR	—	National Foundation for Cancer Research
NFID	—	National Foundation for Infectious Diseases
NHGRI	—	Natl. Human Genome Res. Inst.
NICHD	—	National Institute of Child Health and Human Development
NIGMS	—	National Institute of General Medical Sciences
NIH	—	National Institutes of Health
NIGMS	—	Natl. Institute of General Medical Sciences
NIST	—	Natl. Inst. of Standards and Technology
NLGLP	—	National Laboratory Gene Library Project
NLM	—	National Library of Medicine
NMHCC	—	National Managed Health Care Congress
NORD	—	National Organization for Rare Disorders
NRC	—	National Research Council
NRSA	—	National Research Service Award
NSF	—	Natl. Sci. Foundation
NSGC	—	National Society of Genetic Counselors
NSTA	—	National Science Teachers Association
NYAS	—	New York Academy of Science

OAINN — Ohio Aerospace Institute Neural Networks
OBER — Office of Biological and Environmental Research
OHER — Office of Health and Environmental Research
OMIM — Online *Mendelian Inheritance in Man*
OPRR — Office of Protection from Res. Risks
ORAU — Oak Ridge Associated Universities
ORISE — Oak Ridge Inst. for Science and Education
ORF — open reading frame
ORNL — Oak Ridge National Laboratory
OSU — Oregon State University
OTA — Office of Technology Assessment
PACHG — Program Advisory Committee on the Human Genome
PCR — Polymerase Chain Reaction
PFGE — Pulsed-field gel electrophresis
PG — Plant Genome
PIR — Protein Info. Resource
PNNL — Pacific Northwest National Laboratory
PRIM&R — Public Responsibility in Medicine & Research
PSC — Pittsburgh Supercomputing Center
QDFM — Quantitative DNA Fiber Mapping
RAPD — Random Amplified Polymorphic DNA
RARA — Retinoic Acid Receptor

RECOMB	—	Conference on Computational Molecular Biology
RFLP	—	Restriction Fragment Length Polymorphism
RH	—	Radiation Hybrid
RLGS	—	Restriction Landmark Genomic Scanning
SASE	—	Sample Sequencing
SBH	—	Sequencing by Hybridization
SBIR	—	Small Business Innovation Research
SCAN	—	Sequence Comparison ANalysis Program
SCE	—	School of Continuing Education
SCI	—	Society of Chemical Industry
SCW	—	Single-chromosome Workshop
SDC	—	San Diego Conference
SERGG	—	Southeast Regional Genetics Group
SHOM	—	Sequencing by Hybridization on Matrices
SIAM	—	Society for Industrial and Applied Mathematics
SIB	—	Society for Industrial Biology
SIM	—	Society for Industrial Microbiology
SIMS	—	Societal Institute of the Mathematical Sciences
SIVB	—	Society for In Vitro Biology
SNL	—	Sandia National Laboratory
SNP	—	Single Nucleotide Polymorphism
snRNP	—	Small Nuclear Ribonucleo Protein Particle
SQL	—	Standard Query Language

SSLP	—	Single-sequence Length Polymorphism
SSR	—	Simple Sequence Repeat
STM	—	Scanning Tunneling Microscope
STRP	—	Short Tandem Repeat Polymorphism
STS	—	Sequence-tagged Site
SULS	—	Suffolk Univ. Law School
TCR	—	T-cell receptor
TIGR/NIST	—	The Inst. for Genomic Res./Natl. Inst. of Standards and Technol.
UC	—	University of California
UCB	—	University of California, Berkeley
UCLA	—	University of California, Los Angeles
UICM	—	Univ. of Illinois College of Medicine
UMBC	—	University of Maryland, Baltimore County
UNC	—	University of North Carolina at Chapel Hill
UMDS	—	United Medical & Dental Schools (Univ. of London)
UNESCO	—	United Nations Educational, Scientific, and Cultural Organization
USDA	—	U.S. Dept. of Agriculture
UT	—	University of Tennessee
UW	—	University of Washington
WCCS	—	Workshop on Complete cDNA Sequencing
WHS	—	Wolf-Hirschhorn syndrome
WLMG	—	Wellcome Laboratory for Molecular Genetics

WSES	—	World Scientific and Engineering Society
WWW	—	World Wide Web
XIST	—	candidate gene for X-inactivation center
YAC	—	yeast artificial chromosome

List of Glossary of Terms

Activation energy: Extra energy that must be possessed by atoms or molecules in addition to their ground-state energy in order to undergo a particular chemical reaction. Distinct states correspond to minima of a potential energy surface in a configuration space. In this classical picture, the activation energy for transforming state A into state B is the maximum increase in energy (relative to the ground state of A) encountered on a minimum-energy path from A to B. Energy here refers to potential energy; an analogous definition based on free energy can be constructed. When tunneling is considered, lower energy paths become possible, but an activation energy can be associated with the reaction (at a given temperature) via the relationship between temperature and reaction rate.

Adaptive radiation: The evolution of new species or sub-species to fill unoccupied ecological niches.

Aerobe: A microorganism that grows in the presence of oxygen.

Agarose gel electrophoresis: A matrix composed of a highly purified form of agar that is used to separate larger DNA and RNA molecules ranging 20,000 nucleotides.

Agrobacterium: Agrobacteria is a soil bacteria that works as a natural genetic engineer. In nature, it forces plants to become good hosts by inserting genes into plant cells causing plants to make metabolites agrobacterium need for growth. Genetic engineers have taken advantage of agrobacteria's natural abilities by utilizing them to insert an altered piece of DNA into a host organism.

Agrobacterium Tumefaciens: The bacterium which causes crown gall of dicot plants. It inserts its own *Ti* plasmid DNA into the host plant DNA. The inserted DNA produces growth hormones which result in the tumor and provides a habitat for the bacteria. This is an example of natural genetic engineering. The *Ti* plasmid can be used as a transformation vector.

Algorithm: A set of well-defined mathematical rules or operations for solving a problem in a finite number of steps.

Alkane: (1) A saturated, acyclic hydrocarbon structure; usually quite inert. (2) A hydrocarbon containing a double bond; often rather reactive.

Alleles: Alternate forms of a gene or DNA sequence, which occur on either of two homologous chromosomes in a diploid organism.

Alternative mRNA splicing: The inclusion or exclusion of different exons to form different mRNA transcripts.

Amino acid: Any of 20 basic building blocks of proteins—composed of a free amino (NH2) end, a free carboxyl (COOH) end, and a side group (R).

Ampicillin (beta-lactamase): An antibiotic derived from penicillin that prevents bacterial growth by interfering with cell wall synthesis.

Amplify: To increase the number of copies of a DNA sequence, in vivo by inserting into a cloning vector that replicates within a host cell, or in vitro by polymerase chain reaction (PCR).

Anaerobe: An organism that grows in the absence of oxygen.

Analyte: A chemical species targeted for qualitative or quantitative analysis.

Anneal: The pairing of complementary DNA or RNA sequences, via hydrogen bonding, to form a double-stranded polynucleotide. Most often used to describe the binding of a short primer or probe.

Antibiotic resistance: The ability of a microorganism to produce a protein that disables an antibiotic or prevents transport of the antibiotic into the cell.

Antibiotic: A class of natural and synthetic compounds that inhibit the growth of or kill other microorganisms.

Antibody: An immunoglobulin protein produced by B-lymphocytes of the immune system that binds to a specific antigen molecule.

Anticodon: A nucleotide base triplet in a transfer RNA molecule that pairs with a complementary base triplet, or codon, in a messenger RNA molecule.

Antigen: Any foreign substance, such as a virus, bacterium, or protein, that elicits an immune response by stimulating the production of antibodies.

Antigenic determinant: A surface feature of a microorganism or macromolecule, such as a glycoprotein, that elicits an immune response.

Antigenic switching: The altering of a microorganism's surface antigens through genetic rearrangement, to elude detection by the host's immune system.

Antimicrobial agent: Any chemical or biological agent that harms the growth of microorganisms.

Anti-oncogene:

Antisense DNA: DNA strand that is complementary (opposite) to the functional gene. Antisense DNA can block the function of the normal sense DNA. This method was used in the construction of the Flavr Savr tomato.

Antisense RNA: A complementary RNA sequence that binds to a naturally occurring (sense) mRNA molecule, thus blocking its translation.

Aromatic: A term used to describe cyclic pi-bonded structures of special stability.

Asexual reproduction: Nonsexual means of reproduction which can include grafting and budding.

Autoproductivity: The ability of a system, under external control, to automatically produce an identical copy of itself.

Autosome: A chromosome that is not involved in sex determination.

Bacillus thuringiensis (Bt): A bacterium that kills insects; a major component of the microbial pesticide industry. A group of bacteria naturally occurring in soil that produce proteins, when ingested by specific insects, bind to their stomach lining (epithelium cell) receptors, and are toxic. There are thousands of strains of *Bacillus thuringiensis* each toxic to a limited group of insects such as beetles, flies, and moths. The advantage of crops engineered to produce Bt proteins is that they are toxic to a limited number of insects and are harmless to all mammals.

Backcross: Crossing an organism with one of its parent organisms.

Bacteriocide: A class of antibiotics that kills bacterial cells.

Bacteriophage (phage or phage particle): A virus that in- fects bacteria. Altered forms are used as vectors for cloning DNA.

Bacteriophage (phage): Any virus that infects bacteria. They were the first organisms used for the study of molecular genetics and are now widely used as cloning vectors. (From Greek *phagein*, to eat.)

Bacteriorhodopsin: Pigmented protein found in the plasma membrane of a salt-loving bacterium, *Halobacterium halobium*; it pumps protons out of the cell in response to light.

Bacteriostat: A class of antibiotics that prevents growth of bacterial cells.

Bacterium (plural bacteria): Common name for any member of the diverse group of procaryotic organisms. Most are single cells, but multicellular forms also exist.

Bacterium: A single-celled, microscopic prokaryotic organism: a single cell organism without a distinct nucleus.

Band: In gels and blots, bands are visible indications of a particular fragment of a certain size. There may be one to many bands per lane.

Banoprism: Scientists at Northwestern University have created a nanoparticle with a new shape that could be a useful tool in the race to detect biological threats. The nanoprism, which resembles a tiny Dorito, exhibits unusual optical properties that could be used to improve biodetectors, allowing them to test for a far greater number of biological warfare agents or diseases at one time.

Banoscience: The study of phenomena and manipulation of materials at atomic, molecular and macromolecular scales, where properties differ significantly from those at a larger scale. Nanoscience is primarily the extension of existing sciences into the realms of the extremely small (*nanomaterials, nanochemistry, nanobio, nanophysics*, etc.) while *nanoengineering* represents the extension of the engineering fields into the nano-scale realm (*nanofabrication, nanodevices*, etc.). The exponential growth of nanoscience is largely due to the development of new instruments and related techniques that are used to "routinely" probe and manipulate material at the atomic and molecular level. *Scanning probe microscopies,* analytical electron- beam techniques, epitaxial growth facilities, and *synchrotron* radiation sources are all opening huge opportunities.

Banoshells: Procedures that target cancer cells while leaving normal cells untouched, patients controlling the release of medicine in their bodies with an infrared light, and medical test results produced in seconds rather than days – these are three new medical technologies currently being tested by nanotechnology researchers at Rice University. ... The research focuses on nanoshells, a new type of nanoparticle invented by Naomi Halas, Rice professor of electrical and computer engineering and chemistry. Nanoshells are layered nanoparticles whose ability to manipulate light and color can be designed into the nanoparticle by varying the thickness of the nanoparticle's layers.

Basal body: Short cylindrical array of microtubules plus their associated proteins found at the base of a eucaryotic cell cilium or flagellum. Serves as a nucleation site for the growth of the axoneme. Closely similar in structure to a centriole.

Basal lamina (plural basal laminae): Thin mat of extracellular matrix that separates epithelial sheets, and many types of cells such as muscle cells or fat cells, from connective tissue. Sometimes called a basement membrane.

Base pair (bp): A pair of complementary nitrogenous bases in a DNA molecule—adenine-thymine and guanine-cytosine. Also, the unit of measurement for DNA sequences. In DNA, there are four possible bases: cytosine (C), guanine (G), adenine (A), and thymine (T). Cytosine and thymine are pyrimidine bases; adenine and guanine are purine bases. Cytosine is complementary to guanine while adenine is complementary to thymine. If one strand of DNA has the sequence ATTGC then the complementary strand will be TAACG. Two complementary bases constitute a base pair. In RNA, thymine is replaced by uracil.

Beta-DNA: The normal form of DNA found in biological systems, which exists as a right-handed helix.

Beta-Lactamase: Ampicillin resistance gene.

Binding energy: The reduction in the free energy of a system that occurs when a ligand binds to a receptor. Generally used to describe the total energy required to remove something, or to take a system apart into its constituent particles—for example, to separate two atoms from one another, or to separate an atom into electrons and nuclei.

Bio-assemblies or Biomolecular Assemblies: Containing several protein units, DNA loops, lipids, various ligands, etc.

Bioaugmentation: Increasing the activity of bacteria that decompose pollutants; a technique used in bioremediation.

Biochauvinism: The prejudice that biological systems have an intrinsic superiority that will always give them a monopoly on self-reproduction and intelligence.

Biocompatible coated materials: Biocompatible materials usually used in dental and bone implants that enhance biologic fixation, thereby increasing the bond strength between the coated material and bone, and minimize possible biological effects that may result from the implant itself. MeSH, 1999

Biocompatible materials: Synthetic or natural materials, other than drugs, that are used to replace or repair any body tissue or bodily function. MeSH, 1973

Biodefense: The need for improved national defenses against biological attacks are urgently needed. A real-time reporting system needs to be developed, whereby officials can be informed about an emerging threat before an agent has a chance to affect thousands of people. The development of integrated systems for detecting and monitoring biological agents is instrumental to this goal. The development of vaccines and novel detection platforms will enable preparedness, while also benefiting human health.

Biodiversity: The wide diversity and interrelatedness of earth organisms based on genetic and environmental factors.

Biodynotics Biolcgically Inspired Multifunctional Dynamic Robotics: A multidisciplinary, multi-pronged approach with far reaching impact on robotic capabilities for national security applications. Biologically Inspired Multifunctional Dynamic Robotics: BIODYNOTICS will explore the following areas: DYNAMIC MOBILITY, BEHAVIOR, INTEGRATION Biological Sciences, Defense Sciences Office, DARPA.

Bioengineering: Is rooted in physics, mathematics, chemistry, biology, and the life sciences. It is the application of a systematic, quantitative, and integrative way of thinking about and approaching the solutions of problems important to biology, medical research, clinical proactive, and population studies. The NIH Bioengineering Consortium agreed on the following definition for bioengineering research on biology, medicine, behavior, or health recognizing that no definition could completely eliminate overlap with other research disciplines or preclude variations in interpretation by different individuals and organizations. Integrates physical, chemical, or mathematical sciences and engineering principles for the study of biology, medicine, behavior, or health. It advances fundamental concepts, creates knowledge for the molecular to the organ systems levels, and develops innovative biologics, materials, processes, implants, devices, and informatics approaches for the prevention, diagnosis, and treatment of disease, for patient rehabilitation, and for improving health.

Bioenrichment: Adding nutrients or oxygen to increase microbial breakdown of pollutants.

Biofabrication: Includes nanoparticle delivery systems, biomaterials, tissue engineering, implants and prostheses. Using biological processes to synthesize and manufacture chemicals and materials of high value to the Department of Defense. Biological processes are characterized by: low energy barriers (~10Kcal, high temperatures/pressures not required); high reaction-, regio- and stereo- specificity (protection/ deprotection wastes and catalyst poisoning avoided); spatio-temporal control of materials synthesis of defined composition and size with angstrom- level precision; and local control of the dielectric environment (largely eliminating the need for toxic solvents). Potential target materials and chemicals include composites with enhanced mechanical properties, ultra-low k dielectrics and thermoelectrics, optoelectronic materials, photonic devices (waveguides and logic elements), electronic materials (e.g., GaN, InGaN, AlGaN), elastomers, energetic materials, and adhesives. Biofabrication, DARPA.

Biologics: Agents, such as vaccines, that give immunity to diseases or harmful biotic stresses.

Biomass: The total dry weight of all organisms in a particular sample, population, or area.

Biomaterials: Synthetic or natural materials that can replace or augment tissues, organs or body functions.

Biomechanics: Mechanical structures of living organisms (especially muscles and bones).

Biomedical Nanotechnology: (1) the comprehensive monitoring, control, construction, repair, defense, and improvement of all human biological systems, working from the molecular level, using engineered nanodevices and nanostructures; (2) the science and technology of diagnosing, treating, and preventing disease and traumatic injury, of relieving pain, and of preserving and improving human health, using molecular tools and molecular knowledge of the human body; (3) the employment of molecular machine systems to address medical problems, using molecular knowledge to maintain and improve human health at the molecular scale. (4) A rapidly expanding field that includes many potential technologies and approaches. The key to this definition is that phenomena and materials at the nanometer scale are known to have properties that are uniquely attributable to that scale length. Nanomedicine could similarly be defined [as nanotechnology] as the design, synthesis, or application of materials, devices, or technologies in the nanometer-scale for the basic understanding, diagnosis, and / or treatment of disease. Canadian Institute of Health Research, Regenerative Medicine and Nanomedicine, RFA, 2003. The monitoring, repair, construction and control of human biological systems at the molecular level, using engineered nanodevices and nanostructures. (5) the comprehensive monitoring, control, construction, repair, defense, and improvement of all human biological systems, working from the molecular level, using engineered nanodevices and nanostructures; (6) the science and technology of diagnosing, treating, and preventing disease and traumatic injury, of relieving pain, and of preserving and improving human health, using molecular tools and molecular knowledge of the human body; (7) the employment of molecular machine systems to address medical problems, using molecular knowledge to maintain and improve human health at the molecular scale. The use of molecular-scale devices (nanotechnology) to repair damage and boost the immune system. Vapour grown carbon fibres are obtained as shown in this SEM image. The diameters of these fibres can vary from 100 nm to 500 nm. As part of the National Institutes of Health (NIH) Roadmap for Medical Research [nihroadmap.nih. gov], the NIH [nih.gov] will establish a handful of nanomedicine centers. These centers will be staffed by a highly interdisciplinary scientific crew including biologists, physicians, mathematicians, engineers and computer scientists. Research conducted over the first few years will be spent gathering extensive information about how molecular machines are built. A key activity during this time will be the development of a new kind of vocabulary, or lexicon, to define biological parts and processes in engineering terms. Once researchers have completely catalogued the interactions between and within molecules, they can begin to look for patterns and a higher order of connectedness than is possible to identify with current experimental methods. Mapping these networks and understanding how they change over time will be a crucial step toward helping scientists understand nature's rules of biological design. Understanding these rules will, in many years' time, enable researchers to use this information to address biological issues in unhealthy cells.

Biomedical polymers: More and more therapeutic problems are relevant to the use of polymer- based therapeutic aids for a limited period of time, namely the healing time related to the outstanding capacity of living systems to self- repair ... After healing the remaining prosthetic materials or devices become foreign residues or wastes that have to be eliminated from the body. Nowadays, biocompatible polymers that can degrade in the body are developed. The degradation and the elimination of degradation by-products depend on rather complex phenomena that are presently reflected inconsistently by terms issued from the tradition because each domain has developed its own terminology almost independently. This is a source of misunderstandings, confusions and misperceptions among scientists, surgeons, pharmacists, journalists and politicians, the situation being increased by the introduction of degradable polymers in plastic waste management and environmental protection. Therefore, it is urgent to reflect the various phenomena by specific terms, harmonize and enforce their use by the people active in the biomedical, pharmacological and environmental fields, and, last but not least by the publishing media and journalists. IUPAC, Terminology for biomedical (therapeutic) polymers.

BioMEMS Biological MicroElectro Mechanical Systems: Highlights the technical advances in the field that are leading a revolution in medicine, and creating a new generation of analytical devices for medical diagnostics. The meeting will encompass technology developments in micro & nano drug delivery, interface of nanotech and tissue engineering, microfluidics, and miniaturized total analysis systems (microTAS), biosensors, innovations in mass spec, and nanoscale imaging *BioMEMS and Nanotech World*, Aug. 16- 17, 2004, Washington DC

BioMEMS: MEMS used in medicine, that use microchips.

Biomimetic Chemistry: Knowledge of biochemistry, analytical chemistry, polymer science, and biomimetic chemistry is linked and applied to research in designing new molecules, molecular assemblies, and macromolecules having biomimetic functions. These new bio-related materials of high performance, including, for example, enzyme models, synthetic cell membranes, and biodegradable polymers, are prepared, tested, and constantly improved in this division for industrial scale production.

Biomimetic materials: Materials fabricated by BIOMIMETICS techniques, i.e., based on natural processes found in biological systems. [MeSH 2003] (2) Materials that imitate, copy, or learn from nature.

Biomimetic: Imitating, copying, or learning from nature. Nanotechnology already exists in nature; thus, nanoscientists have a wide variety of components and tricks already available. An interdisciplinary field in materials science, ENGINEERING, and BIOLOGY, studying the use of biological principles for synthesis or fabrication of BIOMIMETIC MATERIALS. There is a need to develop the next generation of restorative materials and medical implants. New avenues of scientific inquiry may enable the development of biomaterials that are safe, reliable, "smart", long-lasting, and perform ideally in their respective biological environments.... Over the last few years biomimetics and tissue engineering have emerged as a new vision in the field of tissue and organ repair and restoration. Biomimetics and tissue engineering are interdisciplinary fields that combine information from the study of biological structures and their functions with physics, mathematics, chemistry and engineering for the generation of new materials, tissues and organs. In the area of craniofacial, oral and dental principles from biomimetics and tissue engineering are applied to developing dental and facial implants, new polymers for guided tissue regeneration used in treating periodontal disease and bone and connective tissue defects, coral- based hydroxyapatite replicas for reconstruction of alveolar ridges and other osseous defects, temporomandibular joint (TMJ) and other joint prostheses, formation of bone matrix substitutes, and artificial replicas of bone, skin, and mucosa. The term biomimetics was coined in 1972 in the context of artificial enzymes; it might be defined broadly as "the abstraction of good design from nature". It is a fact of everyday life that nature has managed to built materials and 'devices' with breathtaking functionality, heterogeneity and stability by using a comparatively limited number of building blocks (the whole range of synthetic materials is restricted to man- made engineering). The basic concepts of nature are often simple; it is the way in which building blocks and materials are arranged that results in functionality. Among the most simple and abundant themes of nature is self- assembly: lipids assemble in sheets to form cell membranes, proteins assemble into functional enzymes, cellular 'sensors', fibers, or virus coats, and DNA assembles in double strands to provide the very basis for live: replication. Biomimetics Study of the structure and function of biological substances to make artificial products that mimic the natural ones.

BIOML Biopolymer Markup Language: Designed by the BIOML core team at Proteometrics, LLC and Proteometrics Canada Ltd. It is to be used "for the annotation of biopolymer sequence information. BIOML allows the full specification of all experimental information known about molecular entities composed of biopolymers, for example, proteins and genes."

Biomolecular materials: An emerging discipline, materials whose properties are abstracted from biology. They share many of the characteristics of biological materials but are not necessarily of biological origin. For example, they may be inorganic materials that are organized or processed in a biomimetic fashion. A key feature of biological and biomolecular materials is their ability to undergo *self- assembly*.

Biomolecular Nanotechnology: Nanotechnology existing in living systems and resulting from our ability to use biomolecules as components for molecular nanotechnology.

Biomotors: Driven by energy sources such as *adenosine triphosphate (ATP)* for chemical transduction and other processes. These biomotors are considered to be biomolecular and are discussed in the body of this report, but strictly speaking they do not conform to the panel's definition of self- assembly.

Bionanotechnology: Includes molecular motors, biomaterials, single molecule manipulation technologies, biochip technologies, etc.

Biopolymeroptoelectromechanical Systems [BioPOEMS]: Combining optics and microelectromechanical systems, and used in biological applications.

Biopolymers: Macromolecules (including proteins, nucleic acids and polysaccharides) formed by living organisms.

Bioremediation: The use of microorganisms to remedy environmental problems.

Biorobotics: Our research focuses on the role of sensing and mechanical design in motor control, in both robots and humans. This work draws upon diverse disciplines, including biomechanics, systems analysis, and neurophysiology. The main approach is experimental, although analysis and simulation play important parts. In conjunction with industrial partners, we are developing applications of this research in biomedical instrumentation, teleoperated robots, and intelligent sensors.

Biosensor: The term "biosensor" is a general designation that denotes either a sensor to detect a biological substance or a sensor which incorporates the use of biological molecules such as antibodies or enzymes. Biosensors are a subcategory of chemical sensors.

Biostasis: A condition in which an organism's cell and tissue structure are preserved, allowing later restoration by cell repair machines. Applicable to cryonics.

Biotechnology: The scientific manipulation of living organisms, especially at the molecular genetic level, to produce useful products. Gene splicing and use of recombinant DNA (rDNA) are major techniques used. Technology for working with biological systems. Includes genetic engineering, human and veterinary medicine, crop and animal breeding, diagnostics, pharmaceuticals, forensics, etc. Narrow sense: Genetic engineering.

Biotic stress: Living organisms which can harm plants , such as viruses, fungi, and bacteria, and harmful insects.

Biovorous: From "biovore;" an organism capable of converting biological material into energy for sustenance.

Bipolar-junction transistor: Transistor with n-type and p-type semiconductors having base-emitter and collector-base junctions.

Blocks Substitution Matrix: A substitution matrix in which scores for each position are derived from *observations* of the frequencies of substitutions in blocks of local alignments in related proteins. Each matrix is tailored to a particular evolutionary distance. In the BLOSUM62 matrix, for example, the alignment from which scores were derived was created using sequences sharing no more than 62% identity. Sequences more identical than 62% are represented by a single sequence in the alignment so as to avoid over-weighting closely related family members. (Henikoff and Henikoff)

Blot, northern: A pattern of RNA fragments transferred to a nitrocellulose membrane from a gel. The gel has undergone electrophoresis to separate fragments according to size. The RNA fragments are arranged in different lanes for each sample. Each lane contains bands which are fragments of different sizes. Radioactive probes are often used to visualized particular bands by autoradiography.

Blot, southern: A pattern of DNA fragments transferred to a nitrocellulose membrane from a gel. The gel has undergone electrophoresis to separate fragments according to size. The DNA fragments are arranged in different lanes for each sample. Each lane contains bands which are fragments of different sizes. Radioactive probes are often used to visualized particular bands by autoradiography.

Blot, western: A pattern of proteins transferred to a nitrocellulose membrane from a gel. The gel has undergone electrophoresis to separate fragments according to size. The proteins are arranged in different lanes for each sample. Each lane contains bands which are fragments of different sizes. Labeled antibody probes are often used to visualize particular bands.

Bone substitutes: Synthetic or natural materials for the replacement of bones or bone tissue. They include hard tissue replacement polymers, natural coral, hydroxyapatite, beta- tricalcium phosphate, and various other biomaterials. The bone substitutes as inert materials can be incorporated into surrounding tissue or gradually replaced by original tissue.

Brownian Assembly: Brownian motion in a fluid brings molecules together in various position and orientations. If molecules have suitable complementary surfaces, they can bind, assembling to form a specific structure. Brownian assembly is a less paradoxical name for self-assembly (how can a structure assemble itself, or do anything, when it does not yet exist?).

Bt (*Bacillus thuringiensis*): A soil bacterium that produces insecticidal proteins. There are several different kinds of proteins produced by different strains of Bt. Some are effective against larvae of moths and butterflies. Others are effective against larvae of beetles. The Bt protein has been introduced into various crops as a built-in insecticide.

Calorie: Unit of heat. One calorie (small "c") is the amount of heat needed to raise the temperature of 1 gram of water by 1 °C. A kilocalorie is the unit used to describe the energy content of foods.

CAP (catabolite gene activator protein): Gene regulatory protein in procaryotes that, when glucose is absent, activates genes responsible for the breakdown of alternative carbon sources.

Capacitor: Energy storage circuit element having two conductors separated by an insulator.

Carcinogen: A substance that induces cancer.

Carcinoma: A malignant tumor derived from epithelial tissue, which forms the skin and outer cell layers of internal organs.

Catalyst: A substance that promotes a chemical reaction by lowering the activation energy of a chemical reaction, but which itself remains unaltered at the end of the reaction.

Catalytic antibody (abzyme): An antibody selected for its ability to catalyze a chemical reaction by binding to and stabilizing the transition state intermediate.

Catalytic RNA (ribozyme): A natural or synthetic RNA molecule that cuts an RNA substrate.

Cation: A positively charged ion.

cDNA library: A library composed of complementary copies of cellular mRNAs.

cDNA: DNA synthesized from an RNA template using reverse transcriptase. Complementary DNA to a particular RNA fragment.

Cell pharmacology: Delivery of drugs by medical nanomachines to exact locations in the body.

Cell Repair Machine: A system including nanocomputers and molecular scale sensors and tools, programmed to repair damage to cells and tissues. Molecular and nanoscale machines with sensors, nanocomputers and tools, programmed to detect and repair damage to cells and tissues, which could even report back to and receive instructions from a hunan doctor if needed.

Cell surgery: In medical nanorobotics, modifying cellular structures using medical nanomachines. Modifying cellular structures using medical nanomachines.

Cell: A membrane-bound unit, typically microns in diameter. All plants and animals are made up of one or more cells (trillions, in the case of human beings). In general, each cell of a multicellular organism contains a nucleus holding all of the genetic information of the organism. (2) A small structural unit, surrounded by a membrane, making up living things.

Cellular oncogene (proto-oncogene): A normal gene that when mutated or improperly expressed contributes to the development of cancer.

Centers of origin: Usually the location in the world where the oldest cultivation of a particular crop has been identified.

Central dogma: Francis Crick's seminal concept that in nature genetic information generally flows from DNA to RNA to protein.

Centrifugation: Separating molecules by size or density using centrifugal forces generated by a spinning rotor. G forces of several hundred thousand times gravity are generated in ultracentrifugation.

Centromere: The central portion of the chromosome to which the spindle fibers attach during mitotic and meiotic division.

Chemotherapy: A treatment for cancers that involves administering chemicals toxic to malignant cells.

Chloramphenicol: An antibiotic that interferes with protein synthesis.

Chromatid: Each of the two daughter strands of a duplicated chromosome joined at the centromere during mitosis and meiosis.

Chromosome walking: Working from a flanking DNA marker, overlapping clones are successively identified that span a chromosomal region of interest.

Chromosome: A single DNA molecule, a tightly coiled strant of DNA, condensed into a compact structure in vivo by complexing with accessory histones or histone-like proteins. Chromosomes exist in pairs in higher eukaryotes. A linear structure in the nucleus of plants and animals that is visible in light microscopy when stained. The chromosome is a single, long, linear molecule of DNA and associated proteins. Bacteria have a single circular chromosome; other organisms may have many linear chromosomes.

Cistron: A DNA sequence that codes for a specific polypeptide; a gene.

Clone: An exact genetic replica of a specific gene or an entire organism.

Cloning vector: A DNA molecule capable of autonomous replication within the cloning host cell (e.g *E. coli*). The vectors contain restriction enzyme sites for insertion of foreign DNA. Cloning vectors are derived from bacterial plasmids, bacteriophages, or viruses.

Cloning: The mitotic division of a progenitor cell to give rise to a population of identical daughter cells or clones.

Coat protein (capsid): The coating of a protein that enclosed the nucleic acid core of a virus.

Codon: A group of three nucleotides that specifies addition of one of the 20 amino acids during translation of an mRNA into a polypeptide. Strings of codons form genes and strings of genes form chromosomes. Set of three nucleotides that specify a particular amino acid during protein synthesis.

Coenzyme (cofactor): An organic molecule, such as a vitamin, that binds to an enzyme and is required for its catalytic activity.

Colony: A group of identical cells (clones) derived from a single progenitor cell.

Combinatorial materials design: Uses computing power (sometimes together with massive parallel experimentation) to screen many different materials possibilities to optimize properties for specific applications (e.g., catalysts, drugs, optical materials).

Commensalism: The close association of two or more dissimilar organisms where the association is advantageous to one and doesn't affect the other(s).

Competency: An ephemeral state, induced by treatment with cold cations, during which bacterial cells are capable of uptaking foreign DNA.

Complementary DNA or RNA: The matching strand of a DNA or RNA molecule to which its bases pair.

Complementary nucleotides: Members of the pairs adenine-thymine, adenine-uracil, and guaninecytosine that have the ability to hydrogen bond to one another.

Concatemer: A DNA segment composed of repeated sequences linked end to end.

Configuration space: A mathematical space describing

the three-dimensional configuration of a system of particles (e.g., atoms in a nanomechanical structure) as a single point; the configuration space for an *N* particle system has 3*N* dimensions.

Conformation: The shape of a molecule. A molecular geometry that differs from other geometries chiefly by rotation about single or triple bonds; distinct conformations (termed con formers) are associated with distinct potential wells. Typical biomolecules and products of organic synthesis can interconvert among many conformations; typical diamondoid structures are locked into a single potential well, and thus lack conformational flexibility.

Conjugation: The joining of two bacteria cells when genetic material is transferred from one bacterium to another.

Constitutive promoter: An unregulated promoter that allows for continual transcription of its associated gene.

Contiguous (contig) map: The alignment of sequence data from large, adjacent regions of the genome to produce a continuous nucleotide sequence across a chromosomal region.

Coulomb [C]: Measure of electrical charge: 1 C is an amount of charge equal to that of about 6.24x1018 electrons.

Coupling: A connection between more than one object or energy pathway so that together they function as a single unit.

Cross-hybridization: The hydrogen bonding of a single-stranded DNA sequence that is partially but not entirely complementary to a singlestranded substrate. Often, this involves hybridizing a DNA probe for a specific DNA sequence to the homologous sequences of different species.

Crossing-over: The exchange of DNA sequences between chromatids of homologous chromosomes during meiosis.

Cross-pollination: Fertilization of a plant from a plant with a different genetic makeup.

Crosstalk: Electromagnetic noise transmitted between leads or circuits in close proximity to each other.

Crybiology: The science of biology at low temperatures; research in cryobiology has made possible the freezing and storing of sperm and blood for later use.

Crystal Lattice: The regular three-dimensional pattern of atoms in a crystal.

Crystal structure: For crystalline materials, the manner in which atoms or ions are arrayed in space. It is defined in terms of the unit cell geometry and the atom positions within the unit cell.

Crystallescence: In medical nanorobotics, the crystallization of solid solute that is offloaded by nanorobot sorting rotors at a concentration that exceeds the solvation capacity of the surrounding solvent.

Culture: A particular kind of organism growing in a laboratory medium.

Curie temperature (also Curie point) (Tc): The temperature above which a ferromagnetic or ferrimagnetic material becomes paramagnetic. For iron the Curie point is 760oC and for nickel 356oC.

Current [A]: Measure of rate of flow of electric charge: a one-ampere current is a flow of 1 C of charge per second.

Cutoff: Condition in a diode or bipolar junction transistor in which the potential across a p-n junction prevents current flow.

Cyclic AMP (cyclic adenosine monophosphate): A

second messenger that regulates many intracellular reactions by transducing signals from extracellular growth factors to cellular metabolic pathways.

Cyclic: A structure is termed cyclic if its covalent bonds form one or more rings.

Cytogenetics: Study that relates the appearance and behavior of chromosomes to genetic phenomenon.

Dalton: A unit of measurement equal to the mass of a hydrogen atom, 1.67 x 10E-24 gram/L (Avogadro's number).

Denature: To induce structural alterations that disrupt the biological activity of a molecule. Often refers to breaking hydrogen bonds between base pairs in double-stranded nucleic acid molecules to produce in single-stranded polynucleotides or altering the secondary and tertiary structure of a protein, destroying its activity.

Density gradient centrifugation: High-speed centrifugation in which molecules "float" at a point where their density equals that in a gradient of cesium chloride or sucrose.

Diabetes: A disease associated with the absence or reduced levels of insulin, a hormone essential for the transport of glucose to cells.

Dideoxynucleotide (didN): A deoxynucleotide that lacks a 3' hydroxyl group, and is thus unable to form a 3'-5' phosphodiester bond necessary for chain elongation. Dideoxynucleotides are used in DNA sequencing and the treatment of viral diseases.

Digest: To cut DNA molecules with one or more restriction endonucleases.

Diploid cell: A cell which contains two copies of each chromosome.

Directional cloning: DNA insert and vector molecules are digested with two different restriction enzymes to create noncomplementary sticky ends at either end of each restriction fragment. This allows the insert to be ligated to the vector in a specific orientation and prevents the vector from recircularizing.

DNA: DNA is the recipe for life. DNA is a molecule found in the nucleus of every celi and is made up of 4 nucleotides Adenine (A), Guanine (G), Thymine (T), and Cystosine (C). The order of the nucleotides in the DNA strand holds a code of information for the cell.

DNA (Deoxyribonucleic acid): An organic acid and polymer composed of four nitrogenous bases—adenine, thymine, cytosine, and guanine linked via intervening units of phosphate and the pentose sugar deoxyribose. DNA is the genetic material of most organisms and usually exists as a double-stranded molecule in which two antiparallel strands are held together by hydrogen bonds between adeninethymine and cytosine-guanine.

DNA (Deoxyriobonucleic Acid): DNA molecules are long chains consisting of four kinds of nucleotides; the order of these nucleotides encodes the information needed to construct protein molecules. These in turn make up much of the molecular machinery of the cell. DNA is the genetic material of cells.

DNA Chip: Also: Gene Chip and DNA Microchip. A purpose built microchip used to identify mutations or alterations in a gene's DNA.

DNA diagnosis: The use of DNA polymorphisms to detect the presence of a disease gene.

DNA diagnostics - miniaturization of: In the areas of sample preparation and assay, it is clear that miniaturization is key. To reduce the size of samples by a factor of 10 or greater, barriers in microfluidics, micromachining, robotics, microchemistry, nucleic acid chemistry, and surface chemistry must be overcome. To implement miniaturized protocols accurately and efficiently, substantial automation of the process will be required. In the development of miniaturized systems, it is essential that the system can be adapted for high levels of parallelization. Miniaturization poses significant technological risks. Currently, there exists no universally accepted precedent for the handling, replication, amplification, or cloning of DNA in nanoliter volumes. Due to the size and charge of the DNA molecule, and the relative instability of many of the enzymes involved in the sample preparation processes, nanoliter and less volumes may pose substantial challenges. In addition, interactions of the biological molecules with the surfaces of the reaction chambers must be minimized. For some methodologies, it is not clear what the optimal sample will be, so substantial improvement in DNA fragmentation technologies or DNA cloning vectors may be required for the ultimate efficient application to diagnostics. Improvements in any of these areas are likely to be of value to other non- DNA based diagnostic applications such as antibody screening protocols and enzyme based diagnostics, because miniaturized robotic or micro-electro mechanical systems developed for DNA could be modified to be used for these purposes.

DNA fingerprint: The unique pattern of DNA fragments identified by Southern hybridization (using a probe that binds to a polymorphic region of DNA) or by polymerase chain reaction (using primers flanking the polymorphic region).

DNA polymorphism: One of two or more alternate forms

(alleles) of a chromosomal locus that differ in nucleotide sequence or have variable numbers of repeated nucleotide units.

DNA Probes: A DNA or nucleic acid probe is a short strand of DNA that locates and binds to its complementary sequence in samples containing single strands of DNA or RNA enabling identification of specific sequences. Nucleic acid probe assays exploit the fundamental hybridization reaction that occurs spontaneously between two complementary DNA:DNA or DNA:RNA strands. As in immunoassays, detection of the hybrid requires that the probe be labeled. Various direct and indirect methods have been devised for the detection of the hybrid. Direct labeling involves attaching the label directly to the probe sequence; indirect labeling binds an antibody to the DNA:DNA or DNA:RNA hybrid. As in immunoassays, non-isotopically-labeled probes are preferred over radio-labeled probes primarily because of radiation hazards, disposal problems, and short reagent shelf life. In addition, the factors determining the detection limits of hybridization assays based on labeled probes are similar to those in immunoassays. Therefore, the development of a simple, inexpensive and sensitive direct detection system which eliminates the use of labels is highly desirable.

DNA Sequencing: There are two main classical methods for sequencing DNA: The first method, developed by Allan Maxam and Walter Gilbert, involves chemicals used to cleave the DNA at certain positions, generating a set of fragments that differ by one nucleotide. The second method, developed by Fred Sanger and Alan Coulson, involves enzymatic synthesis of DNA strands that terminate in a modified nucleotide. Analysis of fragments is similar for both methods and involves gel electrophoresis and autoradiography or fluorescence. The enzymatic method has largely replaced the chemical method as the technique of choice, although there are some situations where chemical sequencing can provide data more easily than the enzymatic method.

DNA, recombinant: DNA that has been cut and spliced back together in a new sequence. The DNA may be from one organism or from more than one organism

DNA, repetitive: Fragments of DNA that appear in multiple copies in a single individual.

DNA: A molecule encoding genetic information, found in the cell's nucleus. Deoxyribose Nucleic Acid. DNA is a molecule used in biological systems to store genetic information. It has a long, double stranded structure with the two strands wrapped around each other to form a right handed helix. There are about 10 nucleotide pairs per helical turn. Each strand is composed of a sugar phosphate backbone and attached bases. It is connected to a complementary strand by hydrogen bonding (non- covalent) between paired bases, adenine (A) with thymine (T) and guanine (G) with cytosine (C). DNA is an abbreviation of DeoxyriboNucleic Acid. : It is the genetic 'blueprint' of all life on this planet. : DNA typing is often applied to Human Identification which has also been called DNA Fingerprinting. DNA Fingerprinting has been attributed to Alex Jeffries, who with his colleagues in 1985, discovered a sequence of nucleotides, subsequently termed short tandem repeats (STR) when comparing the DNA sequence of the human á-globin chromosomes isolated from different individuals.

DNA: Deoxyribonucleic acid, the molecule that stores genetic information. Composed of two complementary strands.

Dominant gene: A gene whose phenotype is when it is present in a single copy.

Dominant(-acting) oncogene: A gene that stimulates cell proliferation and contributes to oncogenesis when present in a single copy.

Dominant: An allele is said to be dominant if it expresses its phenotype even in the presence of a recessive allele.

Dormancy: A period in which a plant does not grow, awaiting necessary environmental conditions such as temperature, moisture, nutrient availability.

Double helix: Describes the coiling of the antiparallel strands of the DNA molecule, resembling a spiral staircase in which the paired bases form the steps and the sugar-phosphate backbones form the rails.

Double-stranded complementary DNA (dscDNA): A duplex DNA molecule copied from a cDNA template.

Downstream: The region extending in a 3' direction from a gene.

E: coli (Escherichia coli): A common intestinal bacterium that is widely used in genetic engineering as a host for a cloning vector. Some strains of E. coli are important foodborne pathogens. Lab strains are of the harmless variety.

Ecology: The study of the interactions of organisms with their environment and with each other.

Ecosystem: The organisms in a plant population and the biotic and abiotic factors which impact on them.

Electrophoresis: The technique of separating charged molecules in a matrix to which is applied an electrical field.

Electroporation: A method for transforrning DNA, especially useful for plant cells, in which high voltage pulses of electricity are used to open pores in cell membranes, through which foreign DNA can pass.

Electroporation: Transformation technique that uses electric fields to temporarily increase permeability of cells to foreign DNA.

Emergence: A complex whole created by simple parts, as in the brain where billions of neurons work individually, but collectively make up our consciousness and give us the ability to think, rationalize, and create.

Emulation: An absolutely precise simulation of something, so exact that it is equivalent to the original (for example, many computers emulate obsolete computers to run their programs). [AS] The Star Trek replicator is an example.

Enabling science and technologies: Areas of research relevant to a particular goal, such as nanotechnology. Also, technology that "enables" other technology to advance, such as the transistor enabled the computer chip revolution, as did photolithography.

Encapsidation: Process by which a virus' nucleic acid is enclosed in a capsid.

Endocrine cell: Specialized animal cell that secretes a hormone into the blood; usually part of a gland, such as the thyroid or pituitary gland.

Endocytosis: Uptake of material into a cell by an invagination of the plasma membrane and its internalization in a membrane-bounded vesicle.

Endophyte: An organism that lives inside another.

Endoplasmic reticulum (ER): Labyrinthine, membrane-bounded compartment in the cytoplasm of eucaryotic cells, where lipids are synthesized and membrane-bound proteins are made.

Endosome: Membrane-bounded organelle in animal cells that carries materials newly ingested by endocytosis and passes many of them on to lysosomes foı degradation.

Endothelium: Single sheet of highly flattened cells (endothelial cells) that forms the lining of all blood vessels. Regulates exchanges between the bloodstream and surrounding tissues and is usually surrounded by a basal lamina.

Energy [J]: Capacity for performing work or to cause heat flow. Like work itself, it is measured in Joules.

Environmental Protection Agency (EPA): The U.S. regulatory agency for biotechnology of microbes. The major laws under which the agency has regulatory powers are the Federal Insecticide, Fungicide, and Rodenticide Act (FIFRA); and the Toxic Substances Control Act (TSCA).

Enzymes: Proteins that control the various steps in all chemical reactions.

EPROM: Electrically Programmable Read-Only Memory — nonvolatile memory device.

Escherichia coli: A commensal bacterium inhabiting the human colon that is widely used in biology, both as a simple model of cell biochemical function and as a host for molecular cloning experiments.

Ethidium bromide: A fluorescent dye used to stain DNA and RNA. The dye fluoresces when exposed to UV light.

Eukaryote: An organism whose cells possess a nucleus and other membrane-bound vesicles, including all members of the protist, fungi, plant and animal kingdoms; and excluding viruses, bacteria, and blue-green algae.

Eurisko: A computer program developed by Professor Douglas Lenat which is able to apply heuristic rules for performing various tasks, including the invention of new heuristic rules.

Event: An "event" in genetic engineering is the insertion of a particular piece of foreign DNA into the chromosome of the recipient. Insertion occurs in random locations, so each event is unique. The event can affect how a gene is expressed in the organism. Once an event occurs, the transgene can be passed to the next generation as a normally inherited gene.

Evolution: The long-term process through which a population of organisms accumulats genetic changes that enable its members to successfully adapt to environmental conditions and to better exploit food resources.

Exon: A DNA sequence that is ultimately translated into protein.

Expectation value: The number of different alignents with scores equivalent to or better than S that are expected to occur in a database search by chance. The lower the E value, the more significant the score.

Express: To translate a gene's message into a molecular product.

Face-centered cubic: A crystal structure found in common elemental metals also FCC. Within the cubic unit cell, atoms are located at all corner and face-centered positions.

Fail-stop: Describes a component or subsystem that, in the event of a failure, produces no output (e.g., of material or data) rather than producing a damaged or incorrect output.

Fermi energy: The energy level in a solid at which the probability of finding an electron is 1/2. For a metal, the energy corresponding to the highest filled electron state in the valence band at 0 K.

Field-Effect Transistor (FET): Field-Effect Transistor — semiconductor device whose insulated gate electrode controls current flow.

FIFRA: The Federal Insecticide, Fungicide, and Rodenticide Act.

Fingerprint: A set of molecular markers sufficiently diverse to identify particular individuals with reasonable certainty.

Flanking region: The DNA sequences extending on either side of a specific locus or gene.

Food and Drug Administration (FDA): The U.S. agency responsible for regulation of biotechnology food products. The major laws under which the agency has regulatory powers include the Food, Drug, and Cosmetic Act; and the Public Health Service Act.

Fusion gene: A hybrid gene created by joining portions of two different genes (to produce a new protein) or by joining a gene to a different promoter (to alter or regulate gene transcription).

Gamete: A haploid sex cell, egg or sperm, that contains a single copy of each chromosome.

Ganglioside: Any glycolipid having one or more sialic acid residues in its structure. Found in the plasma membrane of eucaryotic cells and especially abundant in nerve cells.

Gel, agarose: A substance used to separate DNA or RNA fragments by size. Used for southern and northern blots.

Gel, polyacrylamide: A substance used to separate proteins by size. Used for western blots.

GEM: A genetically engineered microorganism.

Gene amplification: The presence of multiple genes. Amplification is one mechanism through which proto-oncogenes are activated in malignant cells.

Gene cloning: The process of synthesizing multiple copies of a particular DNA sequence using a bacteria cell or another organism as a host.

Gene expression: The process of producing a protein from its DNA- and mRNA-coding sequences.

Gene flow: The exchange of genes between different but (usually) related populations.

Gene frequency: The percentage of a given allele in a population of organisms.

Gene gun: Transformation technique that uses accelerated particles coated with DNA to introduce foreign DNA into recipient plant.

Gene insertion: The addition of one or more copies of a normal gene into a defective chromosome.

Gene linkage: The hereditary association of genes located on the same chromosome.

Gene modification: The chemical repair of a gene's defective DNA sequence.

Gene pool: The totality of all alleles of all genes of all individuals in a particular population.

Gene therapy: The introduction of new genes into individuals to cure diseases or genetic abnormalities.

Gene translocation: The movement of a gene fragment from one chromosomal location to another, which often alters or abolishes expression.

Gene: The basic unit of inheritance. A segment of DNA that codes for a particular protein. A locus on a chromosome that encodes a specific protein or several related proteins. It is considered the functional unit of heredity.

Genetic assimilation: Eventual extinction of a natural species as massive pollen flow occurs from another related species and the older crop becomes more like the new crop.

Genetic code: The three-letter code that translates nucleic acid sequence into protein sequence. The relationships between the nucleotide base-pair triplets of a messenger RNA molecule and the 20 amino acids that are the building blocks of proteins.

Genetic disease: A disease that has its origin in changes to the genetic material, DNA. Usually refers to diseases that are inherited in a Mendelian fashion, although noninherited forms of cancer also result from DNA mutation.

Genetic engineering: Genetic engineering is the process of manually adding new DNA on a molecular level with the goal of adding one or more new traits that are not already found in that organism. For the purposes of this web site, genetic engineering includes recombinant DNA and gene splicing technologies.

Genetic linkage map: A linear map of the relative positions of genes along a chromosome. Distances are established by linkage analysis, which determines the frequency at which two gene loci become separated during chromosomal recombination.

Genetic marker: A gene or group of genes used to "mark" or track the action of microbes.

Genetic modification: The selective and purposeful

alteration of genes by humans. Genetic modification includes methods of incorporating new genes or traits into an organism through genetic engineering, mutagenesis and traditional breeding methods.

Genome: The genetic complement contained in the chromosomes of a given organism, usually the haploid chromosome state. The full chromosome set containing all the genes of a particular individual.

Genomic library: A library composed of fragments of genomic DNA.

Genomics: The study of the structure and function of genomes. Genomics usually involves high speed sequencing of the DNA and computer searches for sequences that code for genes.

Genotype: The sum total of the genetic information of an organism including the linkage relationships between genes. The genotype, modified by environment, determines the phenotype.

Genus: A category including closely related species. Interbreeding between organisms within the same category can occur.

Germ cell (germ line) gene therapy: The repair or replacement of a defective gene within the gamete-forming tissues, which produces a heritable change in an organism's genetic constitution.

GMO: Genetically modified organism. An organism that has incorporated a functional foreign gene through recombinant DNA technology. The novel gene exists in all of its cells and is passed through to progeny. Same as transgenic.

Green revolution: Advances in genetics, petrochemicals, and machinery that culminated in a dramatic increase in crop productivity during the third quarter of the 20th century.

Growth factor: A serum protein that stimulates cell division when it binds to its cell-surface receptor.

Growth phase (curve): The characteristic periods in the growth of a bacterial culture, as indicated by the shape of a graph of viable cell number versus time.

GUI: Graphical User Interface — hardware, software, and firmware that produces the display on modern personal computers.

Guy Fawkes Scenario: If nanotechnology becomes widely available, it might become trivial for anyone to committ acts of terrorism (such as making nanomachines build a large amount of explosives under government buildings a la Guy Fawkes). This would either force strict control over nanotechnology (hard) or a decentralized mode of organization.

Hair cell: Specialized sensory epithelial cell in the ear with bundles of giant microviili (stereocilia) protruding from its apical surface. Sound vibrations tilt the stereocilia, evoking an electrical change in the hair cell, which thus acts as a sound detector.

Hall effect: The phenomenon whereby a force is applied to a moving electron or hole by a magnetic field that is applied perpendicular to the direction of motion. The force direction is perpendicular to both the magnetic field and the particle motion directions.

Haploid cell: A cell containing only one set, or half the usual (diploid) number, of chromosomes.

Heme: An iron complex. Cyclic organic molecule containing an iron atom that carries oxygen in hemoglobin and carries an electron in cytochromes.

Hemophilia: An X-linked recessive genetic disease, caused by a mutation in the gene for clotting factor VIII (hemophilia A) or clotting factor IX (hemophilia B), which leads to abnormal blood clotting.

Herbicide resistance: Herbicide resistance is when plants become resistant to the negative effects of a particular herbicide formulation or are genetically engineered in order to have resistance to a particular herbicide. Herbicide-resistant crops can be sprayed with a particular herbicide, killing weeds growing in and around the crop plants, without being damaged.

Herbicide: Any substance that is toxic to plants; usually used to kill specific unwanted plants.

Heterochromatin: Dark-stained regions of chromosomes thought to be for the most part genetically inactive.

Heteroduplex: A double-stranded DNA molecule or DNA-RNA hybrid, where each strand is of a different origin.

Heterogeneous nuclear RNA (hnRNA): The name originally given to large RNA molecules found in the nucleus, which are now known to be unedited mRNA transcripts, or pre-mRNAs.

Homologous chromosomes: Chromosomes that have the same linear arrangement of genes—a pair of matching chromosomes in a diploid organism.

Homologous recombination: The exchange of DNA fragments between two DNA molecules or chromatids of paired chromosomes (during crossing over) at the site of identical nucleotide sequences.

Homozygote: An organism whose genotype is characterized by two identical alleles of a gene.

Host: An organism that contains another organism.

Human Genome Project: A project coordinated by the National Institutes of Health (NIH) and the Department of Energy (DOE) to determine the entire nucleotide sequence of the human chromosomes.

Human growth hormone (HGH, somatotrophin): A protein produced in the pituitary gland that stimulates the liver to produce somatomedins, which stimulate growth of bone and muscle.

Hybrid: The offspring of two parents differing in at least one genetic characteristic (trait). Also, a heteroduplex DNA or DNA-RNA molecule.

Hybridization, nucleic acid: Complementary strands of DNA or RNA will spontaneously match up with each other and bond together under the right conditions. This is called nucleic acid hybridization and is used when probing southern and northern blots.

Hybridization: The hydrogen bonding of complementary DNA and/or RNA sequences to form a duplex molecule.

Hybridoma: A hybrid cell, composed of a B Iymphocyte fused to a tumor cell, which grows indefinitely in tissue culture and is selected for the secretion of a specific antibody of interest.

Hydrogen bond: A relatively weak bond formed between a hydrogen atom (which is covalently bound to a nitrogen or oxygen atom) and a nitrogen or oxygen with an unshared electron pair.

Hydrolysis: A reaction in which a molecule of water is added at the site of cleavage of a molecule to two products.

Imaging Contrast Agent: A molecule or molecular complex that increases the intensity of the signal detected by an imaging technique, including MRI and ultrasound. An MRI contrast agent, for example, might contain gadolinium attached to a targeting antibody. The antibody would bind to a specific target—a metastatic melanoma cell, for example – while the gadolinium would increase the magnetic signal detected by the MRI scanner.

Immortalization: Production of a cell line capable of an unlimited number of cell divisions. Can be the result of a chemical or viral transformation or of fusion with cells of a tumor line.

Immortalizing oncogene: A gene that upon transfection enables a primary cell to grow indefinitely in culture.

In situ: Refers to performing assays or manipulations with intact tissues.

In vivo: Refers to biological processes that take place within a living organism or cell.

Incomplete dominance: A condition where a heterozygous off- spring has a phenotype that is distinctly different from, and intermediate to, the parental phenotypes.

Initiation codon: The mRNA sequence AUG, coding for methionine, which initiates translation of mRNA.

Inositol lipid: A membrane-anchored phospholipid that transduces hormonal signals by stimulating the release of any of several chemical messengers.

insect resistance management: A set of strategies to reduce the frequency of resistance in crop pests. These strategies are widely referred to as Insect Resistance Management.

Insertion mutations: Changes in the base sequence of a DNA molecule resulting from the random integration of DNA from another source.

Insulator: A substance, object or material that does not conduct electricity. Material that conducts electricity very poorly.

Insulin: A peptide hormone secreted by the islets of Langerhans of the pancreas that regulates the level of sugar in the blood.

Integrated Circuit (IC): An electronic circuit consisting of many interconnected devices on one piece of semiconductor, typically into 10 millimeters on a side. ICs are the major building blocks of today's computers. Semiconductor circuit, typically on a very small silicon chip, containing microfabricated transistors, diodes, resistors, capacitors, etc.

Interferon: A family of small proteins that stimulate viral resistance in cells.

Intergenic regions: DNA sequences located between genes that comprise a large percentage of the human genome with no known function.

Introgression: Backcrossing of hybrids of two plant populations to introduce new genes into a wild population.

Intron: A noncoding DNA sequence within a gene that is initially transcribed into messenger RNA but is later snipped out.

Invasiveness: Ability of a plant to spread beyond its introduction site and become established in new locations where it may provide a deliterious effect on organisms already existing there.

Ion: A charged particle.

Isotope: One of two or more forms of an element that have the same number of protons (atomic number) but differing numbers of neutrons (mass numbers). Radioactive isotopes are commonly used to make DNA probes and metabolic tracers.

Joining (J) segment: A small DNA segment that links

genes to yield a functional gene encoding an immunogobulin.

Jupiter-Brain: A posthuman being of extremely high computational power and size. This is the archetypal concentrated intelligence. The term originated due to an idea by Keith Henson that nanomachines could be used to turn the mass of Jupiter into computers running an upgraded version of himself.

Kanamycin: An antibiotic of the aminoglycoside family that poisons translation by binding to the ribosomes.

Karyotype: All of the chromosomes in a cell or an individual organism, visible through a microsope during cell division.

Kb (kilobase): The number of base pairs (in thousands) that denote the size of a nucleic acid fragment.

Kilohertz (kHz): One thousand cycles per second.

Kilojoule: Standard unit of energy equal to 1000 joules, or 0.24 kilocalories.

Lag phase: The initial growth phase, during which cell number remains relatively constant prior to rapid growth.

Lane: On a gel or a blot, a lane belongs to one sample. It contains fragments of different sizes that were separated by gel electrophoresis.

Lawn: A uniform and uninterrupted laver of bacterial growth, in which individual colonies cannot be observed.

Lectin: Protein that binds tightly to a specific sugar. Abundant lectins derived from plant seeds are often used as affinity reagents to purify glycoproteins or to detect them on the surface of cells.

Legume: A member of the pea family that possesses root nodules containing nitrogen-fixing bacteria.

Library, DNA: A large set of clones of DNA fragments from a particular organism. DNA libraries are usually maintained in *E. coli* or in bacteriophages.

Life (lifetime): The length of time the sensor can be used before its performance changes.

Life Science Applications of Deep Red T2-MP EviTag Quantum Dots: The new T2-MP EviTags are available in the deep red wavelengths used in life science research – Macoun Red (660nm) and Jonamac Red (680nm) – with carboxyl, amine, biotin and non-functionalized surfaces. Having a biologically compatible quantum dot label like the T2-MP EviTags allows further exploration and development of in vivo assays, following cells inside living organism to get a new understanding of life. In particular, it offers the possibility to create new classes of assays for use where current testing materials could produce cytotoxicity or cell death. Potential applications include live cell and even whole animal imaging, blood cancer assays, and numerous other applications.

Ligand: In protein chemistry, a small molecule that is (or can be) bound by a larger molecule is termed a ligand. In organometallic chemistry, a moiety bonded to a central metal atom is also termed a ligand; the latter definition is more common in general chemistry.

Ligase (DNA ligase): An enzyme that catalyzes a condensation reaction that links two DNA molecules via the formation of a phosphodiester bond between the 3' hydroxyl and 5' phosphate of adjacent nucleotides.

Ligate: The process of joining two or more DNA fragments.

Ligation: Joining two fragments of DNA end to end.

Light Emitting Diodes (LEDs): Light Emitting Diodes work on a completely different concept. Traditionally

LEDs are created from two semiconductors. By running current in one direction across the semiconductor the LED emits light of a particular frequency (hence a particular color) depending on the physical characteristics of the semiconductor used. The semiconductor is covered with a piece of plastic that focuses the light and increases the brightness. These semiconductors are very durable, there is no filament, they donít require much power, theyíre brighter and they last a long time. By densely packing red, blue and green LEDs next to each other on a substrate one can create a display.

Lineage: A chart that traces the flow of genetic information from generation to generation.

Linkage: The frequency of coinheritance of a pair of genes and/or genetic markers, which provides a measure of their physical proximity to one another on a chromosome. Measures the physical distance between two genes. Genes that are close together are unlikely to segregate in a sexual cross. Distant genes segregate independently are are then said to be unlinked.

Linked genes/markers: Genes and/or markers that are so closely associated on the chromosome that they are coinherited in 80% or more of cases.

Linker: A short, double-stranded oligonucleotide containing a restriction endonuclease recognition site, which is ligated to the ends of a DNA fragment.

Liposome: A type of nanoparticle made of lipids, or fat molecules, surrounding a water core. Liposomes, several of which are widely used to treat infectious diseases and cancer, were the first type of nanoparticle to be used to create therapeutic agents with novel characteristics. Membrane-bound vesicles constructed in the laboratory to transport biological molecules.

Liquid Crystal Display: LCD is the predominant technology used in flat panel displays. The principle that makes the display work is this: A crystalís alignment can be altered with an electric current. If the crystal is lined up one way ñ it will allow the light waves to pass through a polarized filter, but if the electric current alters the crystalís alignment, it will guide light so that the polarized filter blocks the light. By densely packing red, blue and green light emitting crystals next to each other on a sheet (ìcalled a substrateî), one can create a full color display. The great thing about LCD is that the crystals can be packed together closely, allowing for a higher-resolution, finer-detail display. The con is that LCDs are somewhat fragile, require a lot of power and are relatively less bright. Liquid Crystal Display — display device employing light source and electrically alterable optically active thin film.

Locus (plural = loci): A specific location or site on a chromosome.

Logarithmic phase (log or exponential growth phase): The steepest slope of the growth curve—the phase of vigorous growth during which cell number doubles every 20-30 minutes.

Lysogen: A bacterial cell whose chromosome contains integrated viral DNA.

Lysogenic: A type or phase of the virus life cycle during which the virus integrates into the host chromosome of the infected cell, often remaining essentially dormant for some period of time.

Lytic: A phase of the virus life cycle during which the virus replicates within the host cell, releasing a new generation of viruses when the infected cell lyses.

Machine-phase chemistry: The chemistry of systems in which all potentially reactive moieties follow controlled trajectories (e.g., guided by molecular machines working in vacuum).

Macroscale: Larger than nanoscale; often implies a design that humans can directly interact with; too large to be built by a single assembler (one cubic micron of diamond contains 176 billion atoms).

Magnetization (M): The total magnetic moment per unit volume of material. Also, a measure of the contribution to the magnetic flux by some material within an H field. The magnitude of M is proportional to the applied field as: $M = \div m \times H$, with $\div m$ the magnetic susceptibility.

Magnetostrictive material: A material that changes dimension in the presence of a magnetic field or generates a magnetic field when mechanically deformed.

Malignant: Having the properties of cancerous growth.

Manufacturing: The ability to make products, in this case ranging from clothing, to electronics, to medical devices, to books, to building materials, and much more.

Meat Machine: AKA Cabinet Beast. A box containing assemblers and raw material, within which is formed meat [or whatever else it was programmed to make].

Mechanical: Pertaining to the positions and motions of atoms, as defined by the positions of their nuclei; see electronic. A purely mechanical device can be described in terms of atomic positions and motions without reference to electronic properties, save through their effect on the potential energy function.

Mechanochemistry: Chemistry accomplished by mechanical systems directly controlling the reactant molecules;

the formation or breaking of chemical bonds under direct mechanical control. In this volume, the chemistry of processes in which mechanical systems operating with atomic-scale precision either guide, drive, or are driven by chemical transformations. In general usage, the chemistry of processes in which energy is converted from mechanical to chemical form, or vice versa.

Mechanosynthesis: Chemical synthesis controlled by mechanical systems operating with atomic-scale precision, enabling direct positional selection of reaction sites; synthetic applications of mechanochemistry. Suitable mechanical systems include AFM mechanisms, molecular manipulators, and molecular mill systems. Processes that fall outside the intended scope of this definition include reactions guided by the incorporation of reactive moieties into a shared covalent framework (i.e., conventional intramolecular reactions), or by the binding of reagents to enzymes or enzymelike catalysts. Molecular tools with chemically specific tip structures can be used, sequentially, to modify a work piece and build a wide range of molecular structures.

Mechatronics: The study of the melding of AI and electromechanical machines to make machines that are greater than the sum of their parts. The synergistic combination of precision mechanical engineering with electronic control.

Megabase cloning: The cloning of very large DNA fragments.

Megabyte (MB): 220 (= 1,048,576, or about one million) bytes of information.

Megahertz (Mhz): One million cycles per second

Meiosis: The reduction division process by which haploid gametes and spores are formed, consisting of a single duplication of the genetic material followed by two mitotic divisions.

Meme: An idea that replicates through a society as it is propagated through person-to-person interaction, both direct and indirect. Memetics is a field of study that focuses on memes' role in the evolution of a culture. Self-reproducing idea or other information pattern which is propagated in ways similar to that of a gene. [Richard Dawkins, 1976]

Messenger RNA (mRNA): The class of RNA molecules that copies the genetic information from DNA, in the nucleus, and carries it to ribosomes, in the cytoplasm, where it is translated into protein.

Metabolism: The biochemical processes that sustain a living cell or organism.

Metallothionein: A protective protein that binds heavy metals, such as cadmium and lead.

Micellar nanocontainers: Block copolymer micelles are water- soluble biocompatible nanocontainers with great potential for delivering hydrophobic drugs. An understanding of their cellular distribution is essential to achieving selective delivery of drugs at the subcellular level. R. Savic, L. Luo, A. Eisenberg, D. Maysinger, Micellar nanocontainers distribute to defined cytoplasmic organelles Science 300 (5619): 615- 618, Apr. 25, 2003

Micelles: Micelles are small, spherical structures composed of molecules that attract one another to reduce surface tension. The head of the molecule is hydrophilic, meaning it likes water, while the interior portion is hydrophobic, meaning it avoids water.

Microbial mats (biofilms): Layered groups or communities of microbial populations.

Microelectronics: Narrower terms: MEMS, nanoelectronics, optoelectronics, SED Single Electron Devices. Related terms: molecular electronics, semiconductors

Microencapsulation: Individually encapsulated small particles.

Microinjection: A means to introduce a solution of DNA, protein, or other soluble material into a cell using a fine microcapillary pipet.

Microspectrophotometry: Analytical technique for studying substances present at enzyme concentrations in single cells, in situ, by measuring light absorption. Light from a tungsten strip lamp or xenon arc dispersed by a grating monochromator illuminates the optical system of a microscope. The absorbance of light is measured (in nanometers) by comparing the difference between the image of the sample and a reference image. MeSH 1990

Microspheres: Microspheres are generally defined as small spheres made of any material and sized from about 0.5 μm to 100 μm. Similar, but smaller spheres sized 10 to 500 nm are called *nanospheres*. Ideally, microspheres are completely spherical and homogeneous in size, although less perfect particles are often termed microspheres as well. Depending on the preparation method and material used, microspheres show a typical size distribution which often deviates from the monosized ideal.

Microstructure: In material engineering the structural features of a material such as grain boundaries, grain size and structure, subject to observation under a microscope, selective etching etc. In MEMS microstructure unfortunately is also used to designate a micromachined feature. The last decade has seen rapid developments in the fabrication, characterization and conceptual understanding of synthetic microstructures in many different material systems including silicon, III-V and II-VI semiconductors, metals, ceramics and organics. The objective of this journal [Superlattices and Microstructures] is to provide a common interdisciplinary platform for the publication of the latest research results on all such "nanostructures" with dimensions in the range of 1-100 nm; the unifying theme here being the dimensions of these artificial structures rather than the material system in which they are fabricated.

Microsystem: A microscale machine that can sense information from the environment and act accordingly. Outside the U.S., it can also refer to microelectromechanical systems (MEMS). [smalltimes glossary, 2002]

MicroTas, micro Total Analysis Systems, uTAS: Although initial research dates back to the early 1970's, the field of micro- TAS formally started in 1990, when Manz et al described the possibility of creating microsystems that would take care of many or all the traditional analytical steps involved in a biochemical analysis (sample introduction, handling, extraction, purification, concentration, filtration, analysis, detection) .... Micro- TAS offer many advantages over traditional analysis systems. Low power consumption and small reaction volumes, faster analysis, ultrasensitive detection, and minimal human intervention are key parameters in the development of micro- TAS. Most biochemical reactions take place in liquid environments. Hence, the development of MicroTAS is intrinsically linked to the design of liquid handling micro-devices. [Biomedical Applications Group

(GAB) Centro Nacional de Microelectronica (CNM-IMB) Bellaterra, Spain, 2000]

Microtubule: A hollow cylindrical tube used for the transport of materials inside cells.

Miller indices: A set of 3 integers (4 for hexagonal) that designate crystallographic planes, as determined from reciprocals of fractional axial intercepts.

Millimeter: One thousandth of a meter, or about 1/26 of an inch.

Mitosis: The replication of a cell to form two daughter cells with identical sets of chromosomes.

Molecular - Nanotechnology: The technology of precisely-constructed molecular-scale machines; from nanometer: a billionth of a meter. For a good introduction, see What is Nanotechnology? or Ralph C. Merkle's nanotechnology page.

Molecular assembler: A general-purpose device for molecular manufacturing, able to guide chemical reactions by positioning individual molecules to atomic accuracy (e.g. mechanosynthesis) and to construct a wide range of useful and stable molecular structures according to precise specifications. Also known as an assembler, a molecular assembler is a molecular machine that can build a molecular structure from its component building blocks.

Molecular Beam Epitaxy: [MBE] Process used to make compound (multi-layer) semiconductors. Consists of depositing alternating layers of materials, layer by layer, one type after another (such as the semiconductors gallium arsenide and aluminum gallium arsenide).

Molecular biology: The study of the biochemical and molecular interactions within living cells.

Molecular cloning: The biological amplification of a specific DNA sequence through mitotic division of a host cell into which it has been transformed or transfected.

Molecular genetics: The study of the flow and regulation of genetic information between DNA, RNA, and protein molecules.

Molecular markers: Genetic traits which are detectable on gels or blots and can be used to construct genetic maps. Molecular markers usually have no known function. RFLPs are molecular markers.

Molecular mechanics models: Many of the properties of molecular systems are determined by the molecular potential energy function. Molecular mechanics models approximate this function as a sum of 2-atom, 3-atom, and 4-atom terms, each determined by the geometries and bonds of the component atoms. The 2-atom and 3-atom terms describing bonded interactions roughly correspond to linear springs.

Molecular medicine: A variety of pharmaceutical techniques and gene therapies that address specific molecular diseases or molecular defects in biological systems. A variety of pharmaceutical techniques and therapies in use today. Studying molecules as they relate to health and disease, and manipulating those molecules to improve the diagnosis, prevention, and treatment of disease.

Molecular mill: A mechanochemical processing system characterized by limited motions and repetitive operations without programmable flexibility.

Molecular motors: Protein based machines that are involved in or cause movement such as the rotary devices (flagellar motor and the F1 ATPase) or the devices whose movement is directed along cytoskeletal filaments (myosin, kinesin and dynein motor families).

Molecular Nanoscience: An emerging interdisciplinary field that combines the study of molecular/ biomolecular systems with the science and technology of nanoscale structures and systems. The potential applications for this research are very broad and include such possibilities as 1) the use of biomolecules and cellular systems to self- assemble nanoelectronic circuitry and other nanoscale structures and 2) the use of lamellar host frameworks containing nano pores that can be tailored to include guest molecules for separation of chemicals for pharmaceutical and other applications. Molecular Nanoscience Alliance for Interdisciplinary Studies and Activities, Univ. of Minnesota, US,

Molecular Wire: A molecular wire—the simplest electronic component—is a quasi-one-dimensional molecule that can transport charge carriers (electrons or holes) between its ends. [Michael D Ward]

Molecularly imprinted polymers MIPs: A new class of materials that have artificially created receptor structures. Since their discovery in 1972, MIPs have attracted considerable interest from scientists and 3engineers involved with the development of chromatographic absorbents, membranes, sensors and enzyme and receptor mimics.

Monoclonal antibodies: Immunoglobulin molecules of single- epitope specificity that are secreted by a clone of B cells.

Monoculture: The agricultural practice of cultivating crops consisting of genetically similar organisms.

Monogenic: Controlled by or associated with a single gene.

Multi-locus probe: A probe that hybridizes to a number of different sites in the genome of an organism.

Mutagen: Any agent or process that can cause mutations.

Mutation: An alteration in DNA structure or sequence of a gene. A spontaneous or induced genetic change in the DNA of an organism. Most mutations are undesirable, but a few are useful such as dwarfing genes in cereal crops that allow the plants to stand better.

Mycorrhizae: Fungi that form symbiotic relationships with roots of more developed plants.

Nanobiology: Many fundamental biological functions are carried out by molecular machineries that have the sizes of 1-100 nm. You find many examples in molecular biology and cell biology: single enzymes, transcription complex, ribosome, transport complex, nuclear pore, and so on. To understand the functions of these machineries, one has to describe their movements, changes in their shapes, and their localization. This means the mechanistic study is equivalent to dynamic morphology at this level of size, making a new field that merges mechanistic biology and morphology. The emergence of nanobiology depended on the invention of nano- technology: scanning probe microscopy, modern optical techniques, and micro- manipulating techniques. This concept of nanobiology was first proposed in a group study named "Biological Nano- Mechanisms", which was supported by Japanese Agency of Science and Technology (1992-1997). [National Institute of Genetics, Japan]

Nanobiotechnology: An emerging area of scientific and technological opportunity. Nanobiotechnology applies the tools and processes of nano/ microfabrication to build devices for studying biosystems. Researchers also learn from biology how to create better micro-nanoscale devices. The Nanobiotechnology Center (NBTC), a National Science Foundation, Science and

Technology Center is characterized by its highly interdisciplinary nature and features a close collaboration between life scientists, physical scientists, and engineers. [NBTC, NanoBiotechnology Center, Cornell Univ. US, 2002] Use of nanotechnology[ies] in the life sciences. These may include, but are not limited to, therapeutics, medical devices/implants, biosensors, and tools for the development of drugs. Applying the tools and processes of MNT to build devices for studying biosystems, in order to learn from biology how to create better nanoscale devices. Should hasten the creation of useful micro devices that mimic living biological systems.

Nanocentrifuge: In medical nanorobotics, a proposed nanodevice that can spin materials at very high speed, imparting rotational accelerations of up to one trillion gravities (g's), thus permitting rapid sortation.

Nanoscience: Nanoscience is the science of nanoscale phenomena. This is a very complicated area, since Quantum Mechanics is often needed to describe the physical properties of atoms and molecules properly. What makes it even more complicated is that nanostructured materials can still be quite a challenge to describe Quantum Mechanically, since the equations are feindishly difficult—this is where supercomputers come in to make the job approachable. The scientific discipline seeking to increase our knowledge and understanding of nanoscale phenomena, i.e. science on the scale of 0.1 nm to 100 nm.

Nanosensor: A chemical or physical sensor constructed using nanoscale components, usually microscopic or submicroscopic in size.

Nanosources: sources that emit light from nanometre-scale volumes.

National Institutions of Health (NIH): A nonregulatory agency which has oversight of research activities that the agency funds.

National Science Foundation (NSF): A nonregulatory agency which has oversight of biotechnology research activities that the agency funds.

Natural selection: The differential survival and reproduction of organisms with genetic characteristics that enable them to better utilize environmental resources.

Neophile: One who welcomes the future and who enjoys change and evolution.

Nick translation: A procedure for making a DNA probe in which a DNA fragment is treated with DNase to produce single-stranded nicks, followed by incorporation of radioactive nucleotides from the nicked sites by DNA polymerase I.

Nicked circle (relaxed circle): During extraction of plasmid DNA from the bacterial cell, one strand of the DNA becomes nicked. This relaxes the torsional strain needed to maintain supercoiling, producing the familiar form of plasmid.

Nitrocellulose: A membrane used to immobilize DNA, RNA, or protein, which can then be probed with a labeled sequence or antibody.

Nitrogen fixation: The conversion of atmospheric nitrogen to biologically usable nitrates.

Nitrogenous bases: The purines (adenine and guanine) and pyrimidines (thymine, cytosine, and uracil) that comprise DNA and RNA molecules.

Nodule: The enlargement or swelling on roots of nitrogen-fixing plants. The nodules contain symbiotic nitrogen-fixing bacteria.

Nontarget organism: An organism which is affected by an interaction for which it was not the intended recipient.

Northern hybridization: (Northern blotting). A procedure in which RNA fragments are transferred from an agarose gel to a nitrocellulose filter, where the RNA is then hybridized to a radioactive probe.

Novel foods: Novel foods are products that have never been used as a food; foods which result from a process that has not previously been used for food; or, foods that have been modified by genetic manipulation. This last category of foods have been described as genetically modified foods (often referred to as GM foods, genetically engineered foods or biotechnology-derived foods).

Nuclease: Any enzyme that cuts nucleic acids. See restriction enzyme. A class of enzymes that degrades DNA and/ or RNA molecules by cleaving the phosphodiester bonds that link adjacent nucleotides. In deoxyribonuclease (DNase), the substrate is DNA. In endonuclease, it cleaves at internal sites in the substrate molecule. Exonuclease progressively cleaves from the end of the substrate molecule. In ribonuclease (RNase), the substrate is RNA. In the S1 nuclease, the substrate is single-stranded DNA or RNA.

Nucleic acids: The two nucleic acids, deoxyribonucleic acid (DNA) and ribonucleic acid (RNA), are made up of long chains of molecules called nucleotides.

Nuclein: The term used by Friedrich Miescher to describe the nuclear material he discovered in 1869, which today is known as DNA.

Nucleoside analog: A synthetic molecule that resembles a naturally occuring nucleoside, but that lacks a bond site needed to link it to an adjacent nucleotide.

Nucleotide: A building block of DNA and RNA, consisting of a nitrogenous base, a five-carbon sugar, and a phosphate group. Together, the nucleotides form codons, which when strung together form genes, which in turn link to form chromosomes.

Nucleus: The membrane-bound region of a eukaryotic cell that contains the chromosomes. A cellular organelle in plants and animals that contains the chromosomes which in turn are composed of DNA plus protein.

Occupational Safety and Health Administration (OSHA): One of the U.S. agencies responsible for regulation of biotechnology. The major law under which the agency has regulatory powers is the Occupational Safety and Health Act.

Oligonucleotide: A DNA polymer composed of only a few nucleotides.

Oncogene: A gene that contributes to cancer formation when mutated or inappropriately expressed.

Oncogenesis: The progression of cytological, genetic, and cellular changes that culminate in a malignant tumor.

Open pollination: Pollination by wind, insects, or other natural mechanisms.

Open reading frame: A long DNA sequence that is uninterrupted by a stop codon and encodes part or all of a protein.

Operator: A prokaryotic regulatory element that interacts with a repressor to control the transcription of adjacent structural genes.

Orbital: In the approximation that each electron in a molecule has a distinct, independent wave function, the spatial distribution of an electron wave function corresponds to a molecular orbital. These, in turn, can be approximated as sums of contributions from the orbitals characteristic of the isolated atoms. An electron added to a molecule—or, similarly, one excited to a higher-energy state within a molecule—would occupy a state with a different wave function from the rest; an unoccupied state of this kind corresponds to an unoccupied molecular orbital. Orbital-symmetry effects on reaction rates arise when a reaction requires overlap between two lobes of the orbitals on each of two reagents: if the algebraic signs of the wave functions in the facing lobes do not match, bond formation between those orbitals is prohibited.

Organelle: A cell structure that carries out a specialized function in the life of a cell.

Origin of replication: The nucleotide sequence at which DNA synthesis is initiated.

Overlapping reading frames: Start codons in different reading frames generate different polypeptides from the same DNA sequence.

Ovum: A female gamete.

Paleontology: The study of the fossil record of past geological periods and of the phylogenetic relationships between ancient and contemporary plant and animal species.

Palindromic sequence: A DNA locus whose 5'-to-3' sequence is identical on each DNA strand. The sequence is the same when one strand is read left to right and the other strand is read right to left. Recognition sites of many restriction enzymes are palindromic.

pAMP: Ampicillin-resistant plasmid developed for this laboratory course.

Parasitism: The closee association of two or more dissimilar organisms where the association is harmful to at least one.

Particle bombardment: Using metal particles to blast DNA into cells for the purpose of genetic engineering. A gene gun is used to propel the metal particles.

Pathogen: Organism which can cause disease in another organism.

pBR322: A derivation of ColE1, one of the first plasmid vectors widely used.

PCR: Polymerase chain reaction. An amazingly sensitive technique that allows the specific amplification of extremely small amounts of particular DNA fragments using DNA polymerase and specific primers.

Pedigree: A diagram mapping the genetic history of a particular family.

Percent Accepted Mutation: A unit introduced by Dayhoff et al. to quantify the amount of evolutionary change in a protein sequence. 1.0 PAM unit, is the amount of evolution which will change, on average, 1% of amino acids in a protein sequence. A PAM(x) substitution matrix is a look-up table in which scores for each amino acid substitution have been calculated based on the frequency of that substitution in closely related proteins that have experienced a certain amount (x) of evolutionary divergence.

Persistence: Ability of an organism to remain in a particular setting for a period of time after it is introduced.

Pesticide: A substance that kills harmful organisms (for example, an insecticide or fungicide).

Phenotype: The observable characteristics of an organism, the expression of gene alleles (genotype) as an observable physical or biochemical trait.

Pheromone: A hormone-like substance that is secreted into the environment.

Phosphatase: An enzyme that hydrolyzes esters of phosphoric acid, removing a phosphate group.

Phosphodiester bond: A bond in which a phosphate group joins adjacent carbons through ester linkages. A condensation reaction between adjacent nucleotides results in a phosphodiester bond between 3' and 5' carbons in DNA and RNA.

Phospholipid: A class of lipid molecules in which a phosphate group is linked to glycerol and two fatty acyl groups. A chief component of biological membranes.

Phosphorylation: The addition of a phosphate group to a compound.

Physical map: A map showing physical locations on a DNA molecule, such as restriction sites, and sequence-tagged sites.

Piezoelectric material: A ferroelectric material in which an electrical potential difference is created due to mechanical deformation, or conversely, in which the application of a voltage causes dimensional changes in the material.

Piezoelectric: A piezoelectric material expands or contracts when a voltage is applied across it. Piezoelectric tubes are often used to move the tip in an AFM or STM. If a piezoelectric crystal is compressed, a voltage appears across it. This effect is sometimes used to measure displacements. A commonly used piezoelectric is PZT (Pb,Zr)TiO3.

Plaque: A clear spot on a lawn of bacteria or cultured cells where cells have been Iysed by viral infection.

Plasmid (p): A circular DNA molecule, capable of

autonomous replication, which typically carries one or more genes encoding antibiotic resistance proteins. Plasmids can transfer genes between bacteria and are important tools of transformation for genetic engineers.

Pleiotrophy: The effect of a particular gene on several different traits.

Point mutation: A change in a single base pair of a DNA sequence in a gene.

Poly(A) polymerase: Catalyzes the addition of adenine residues to the 3' end of pre-mRNAs to form the poly(A) tail.

Polyacrylamide gel electrophoresis: Electrophoresis through a matrix composed of a synthetic polymer, used to separate proteins, small DNA, or RNA molecules of up to 1000 nucleotides. Used in DNA sequencing.

Polyclonal antibodies: A mixture of immunoglobulin molecules secreted against a specific antigen, each recognizing a different epitope.

Polygenic: Controlled by or associated with more than one gene.

Polylinker: A short DNA sequence containing several restriction enzyme recognition sites that is contained in cloning vectors.

Polymer: A molecule composed of repeated subunits.

Polymerase (DNA): Synthesizes a double-stranded DNA molecule using a primer and DNA as a template.

Polymerase chain reaction (PCR): A procedure that enzymatically amplifies a DNA polymerase.

Polymerase: Any enzyme that adds subunits to chains of macromolecules. Example is DNA polymerase.

Polymorphisms: Variant forms of a particular gene that occur simultaneously in a population.

Polynucleotide: A DNA polymer composed of multiple nucleotides.

Polypeptide (protein): A polymer composed of multiple amino acid units linked by peptide bonds.

Polyploid: A multiple of the haploid chromosome number that results from chromosome replication without nuclear division.

Polysaccharide: A polymer composed of multiple units of monosaccharide (simple sugar).

Polyvalent vaccine: A recombinant organism into which has been cloned antigenic determinants from a number of different disease-causing organisms.

Population: A local group of organisms belonging to the same species and capable of interbreeding.

Primary cell: A cell or cell line taken directly from a living organism, which is not immortalized.

Primer: A short DNA or RNA fragment annealed to single-stranded DNA, from which DNA polymerase extends a new DNA strand to produce a duplex molecule.

Probe: A single-stranded DNA that has been radioactively labeled and is used to identify complementary sequences in genes or DNA fragments of interest. A fragment of DNA, usually labeled with radioactive 32P, that detects homologous sequences on southern blots.

Prokaryote: A bacterial cell lacking a true nucleus; its DNA is usually in one long strand.

Promoter: A region of DNA extending 150-300 bp upstream from the transcription start site that contains binding sites for RNA polymerase and a number of proteins that regulate the rate of transcription of the adjacent gene.

Pronucleus: Either of the two haploid gamete nuclei just prior to their fusion in the fertilized ovum.

Protease: An enzyme that cleaves peptide bonds that link amino acids in protein molecules.

Protein: Molecules composed of amino acids. Proteins constitute the enzymes and many of the structural components of cells. A polymer of amino acids linked via peptide bonds and which may be composed of two or more polypeptide chains. Proteins do the work in cells. They can be part of structures (such as cell walls, organelles, etc), regulate reactions that take place in the cell or they can serve as enzymes, which speed-up reactions. Everything you see in an organism is either made of proteins or the result of a protein action.

Protein kinase: An enzyme that adds phosphate groups to a protein molecule at serine, threonine, or tyrosine residues.

Proteinaceous infectious particle (prion): A proposed pathogen composed only of protein with no detectable nucleic acid and which is responsible for Creutzfeldt-Jakob disease and kuru in humans and scrapie in sheep.

Quantum Mechanics: A largely computational physical theory explaining the behavior of quantum phenomena, which incorporates the theory of special relativity. Despite dilignet attempts, general relativity has not been sucessfully incorporated into quantum mechanics. A physical model of chemical and optical phenomena, as well as the behaviour of matter in general on a small scale. Quantum mechanics describes a system of particles in terms of a wave function defined over the configuration space of the system. Although the concept of particles having distinct locations is implicit in the potential energy function that determines the wave function (e.g., of a ground-state system), the observable dynamics of the system cannot be described in terms of the motion of such particles from point to point. In describing the energies, distributions, and behaviors of electrons in nanometer-scale structures, quantum mechanical methods are necessary. Electron wave functions help determine the potential energy surface of a molecular system, which in turn is the basis for classical descriptions of molecular motion. Nanomechanical systems can almost always be described in terms of classical mechanics, with occasional quantum mechanical corrections applied within the framework of a classical model.

Quantum Well: A P-N-P junction in which the "N" layer is ~10 nm (where traditional physics leaves off and quantum effects take over) and an "electron trap" is created. "If one makes a heterostructure with sufficiently thin layers, quantum interference effects begin to appear prominently in the motion of the electrons. The simplest structure in which these may be observed is a quantum well, which simply consists of a thin layer of a narrower-gap semiconductor between thicker layers of a wider-gap material."

rDNA: Def. 1. Ribosomal DNA; DNA which codes for ribosomes

Reading frame: A series of triplet codons beginning from a specific nucleotide. Depending on where one begins, each DNA strand contains three different reading frames.

Recessive gene: Characterized as having a phenotype

expressed only when both copies of the gene are mutated or missing.

Recessive(-acting) oncogene, (anti-oncogene): A single copy of this gene is sufficient to suppress cell proliferation; the loss of both copies of the gene contributes to cancer formation.

Recessive: Refers to the member of a pair of alleles that fails to be expressed in the phenotype of the organism when the dominant member is present. Also refers to the phenotype of an individual that has only the recessive allele.

Recognition sequence (site): A nucleotide sequence—composed typically of 4, 6, or 8 nucleotides—that is recognized by a restriction endonuclease. Type II enzyrnes cut (and their corresponding modification enzymes methylate) within or very near the recognition sequence.

Recombinant DNA: The process of cutting and recombining DNA fragments from different sources as a means to isolate genes or to alter their structure and function.

Recombination frequency: The frequency at which crossing over occurs between two chromosomal loci—the probability that two loci will become unlinked during meiosis.

Refuge: A refuge is any host plant (non-Bt crop, potatoes, oats, sorghum, and some weeds) not producing Bt proteins or not being treated with conventional Bt formulations. The purpose of the refuge is to supply a source of susceptible insects that may mate with resistant insects of the same species emerging from nearby Bt crop fields. If the refuge is large enough, susceptible insect numbers should outnumber resistant individuals, and in theory, a greater number of the next generation offspring will be susceptible to the Bt crop.

Regulatory gene: A gene whose protein controls the activity of other genes or metabolic pathways.

Relaxed plasmid: A plasmid that replicates independently of the main bacterial chromosome and is present in 10-500 copies per cell.

Renature: The reannealing (hydrogen bonding) of single-stranded DNA and/or RNA to form a duplex molecule.

Replicon: A chromosomal region containing the DNA sequences necessary to initiate DNA replication processes.

Repressor: A DNA-binding protein in prokaryotes that blocks gene transcription by binding to the operator.

Resistance: The evolved ability of a species to survive during a stressful event or set of circumstances.

Restriction endonuclease (enzyme): A class of endonucleases that cleaves DNA after recognizing a specific sequence, such as BamH1 (GGATCC), EcoRI (GAATTC), and HindIII (AAGCTT). Type I. Cuts nonspecifically a distance greater than 1000 bp from its recognition sequence and contains both restriction and methylation activities. Type II. Cuts at or near a short, and often symmetrical, recognition sequence. A separate enzyme methylates the same recognition sequence. Type III. Cuts 24-26 bp downstream from a short, asymmetrical recognition sequence. Requires ATP and contains both restriction and methylation activities.

Restriction enzyme: A class of enzymes that cut DNA at specific sequences called restriction sites. Restriction enzymes were key to making genetic engineering possible.

Restriction-fragment-length polymorphism (RFLP): Differences in nucleotide sequence between alleles at a chromosomal locus result in restriction fragments of varying lengths detected by Southern analysis.

Retrovirus: A member of a class of RNA viruses that utilizes the enzyme reverse transcriptase to reverse copy its genome into a DNA intermediate, which integrates into the hostcell chromosome. Many naturally occurring cancers of vertebrate animals are caused by retroviruses.

Reverse genetics: Using linkage analysis and polymorphic markers to isolate a disease gene in the absence of a known metabolic defect, then using the DNA sequence of the cloned gene to predict the amino acid sequence of its encoded protein.

Reverse transcriptase (RNA-dependent DNA polymerase): An enzyme isolated from retrovirus-infected cells that synthesizes a complementary (c)DNA strand from an RNA template.

RFLP: Restriction fragment length polymorphism. RFLPs are generated by digesting DNA with restriction enzymes. The DNA is separated by size by gel electrophoresis. Slight differences in homologous fragments may exist between individuals. These length polymorphisms are RFLPs.

Rhizobia: Bacteria in a symbiotic relationship with leguminous plants that results in nitrogen fixation.

Rhizosphere: The soils region on and around plant roots.

Ribosomal RNA (rRNA): The RNA component of the ribosome.

Ribosome: Cellular organelle that is the site of protein synthesis during translation. See Organelle, Translation.

Ribosome-binding site: The region of an mRNA molecule that binds the ribosome to initiate translation.

Risk assessment: Defined as a formalized basis for the objective evaluation of risk in a manner in which assumptions and uncertainties are clearly considered and presented.

RNA (ribonucleic acid): An organic acid composed of repeating nucleotide units of adenine, guanine, cytosine, and uracil, whose ribose components are linked by phosphodiester bonds.

RNA polymerase: Transcribes RNA from a DNA template.

RNA: Ribonucleic acid. The three basic types are messenger RNA (mRNA), ribosomal RNA (rRNA), and transfer RNA (tRNA).

Roundup Ready: A trademark for plants that are genetically engineered to be resistant to the herbicide Roundup (technical name: glyphosate).

Salmonella: A genus of rod-shaped, gram-negative bacteria that are a common cause of food poisoning.

Satellite RNA (viroids): A small, self-splicing RNA molecule that accompanies several plant viruses, including tobacco ringspot virus.

Scanning Near Field Optical Microscopy: A method for observing local optical properties of a surface that can be smaller than the wavelength of the light used. A type of scanning probe microscopy that maps the near-field optical properties of an interface.

Scanning Probe Microscopy (SPM): A method for imaging nanoscale features of surfaces by scanning a sensor (probe) over a surface. Near-field effects such as tunneling, van der Waals forces, local fields and more are serially detected at localized points on the surface and used to create an SPM image.

Scanning Thermal Microscopy (SThM): A type of scanning probe microscopy that maps the local temperature and thermal conductivity of an interface. A method for observing local temperatures and temperature gradients on a surface. [NTN]

Scanning Tunneling Microscope (STM): An instrument able to image conducting surfaces to atomic accuracy; has been used to pin: molecules to a surface. A device in which a sharp conductive tip is moved across a conductive surface close enough to permit a substantial tunneling current (typically a nanometer or less). In a common mode of operation, the voltage is kept constant and the current is monitored and kept constant by controlling the height of the tip above the surface; the result, under favorable conditions, is an atomic-resolution map of the surface reflecting a combination of topography and electronic properties. The STM has been used to manipulate atoms and molecules on surfaces. A type of scanning probe microscopy that maps the local electrical properties of an interface.

Schwann cell: Glial cell responsible for forming myelin sheaths in the peripheral nervous system.

Science: The process of developing a systematized knowledge of the world through the variation and testing of hypotheses. The pursuit of knowledge and understanding, from the Latin term *scientia*, which means 'knowledge'.

Self-pollination: Pollen of one plant is transferred to the female part of the same plant or another plant with the same genetic makeup.

Semiconservative replication: During DNA duplication, each strand of a parent DNA molecule is a template for the synthesis of its new complementary strand. Thus, one half of a preexisting DNA molecule is conserved during each round of replication.

Sequence hypothesis: Francis Crick's seminal concept that genetic information exists as a linear DNA code; DNA and protein sequence are colinear.

Sequence: Noun: the particular order of nucleotides in a DNA or RNA fragment. Verb: to determine the particular order of nucleotides in a strand of DNA.

Sequence-tagged site (STS): A unique (single-copy) DNA sequence used as a mapping landmark on a chromosome.

Sexual reproduction: The process where two cells (gametes) fuse to form one hybrid, fertilized cell.

Signal transduction: The biochemical events that conduct the signal of a hormone or growth factor from the cell exterior, through the cell membrane, and into the cytoplasm. This involves a number of molecules, including receptors, pro- teins, and messengers.

Site-directed mutagenesis: The process of introducing spe- cific base-pair mutations into a gene.

Small nuclear RNA (snRNA): Short RNA transcripts of 100-300 bp that associate with proteins to form small nuclear ribonucleoprotein particles (snRNPs), which participate in RNA processing.

Somatic cell gene therapy: The repair or replacement of a defective gene within somatic tissue.

Somatic cell: Any nongerm cell that composes the body of an organism and which possesses a set of multiploid chromosomes (diploid in most organisms).

Southern hybridization (Southern blotting): A procedure in which DNA restriction fragments are transferred from an agarose gel to a nitrocellulose filter, where the denatured DNA is then hybridized to a radioactive probe (blotting).

Species: A classification of related organisms that can freely interbreed.

Spore: A form taken by certain microbes that enables them to exist in a dormant stage. It is an asexual reproductive cell.

Stationary phase: The plateau of the growth curve after log growth, during which cell number remains constant. New cells are produced at the same rate as older cells die.

Structure-functionalism: The scientific tradition that stresses the relationship between a physical structure and its function, for example, the related disciplines of anatomy and physiology.

Subcloning: The process of tranferring a cloned DNA fragment from one vector to another.

Substantial equivalence: The concept of substantial equivalence is used as a guide in the safety assessment of genetically modified foods by comparing the novel food to its unmodified counterpart which has a history of safe use. This approach allows regulatory agencies to include in their consideration, the substantial history of information related to foods which have long been safely consumed in the human diet to aid in the identification of potential safety and nutritional issues.

Subunit vaccine: A vaccine composed of a purified antigenic determinant that is separated from the virulent organism.

1

Biotechnology Laboratories: An Overview

Biotechnology is the use of biological systems — living things — to make or change products. It has been used for centuries in traditional activities like baking bread and making cheese. Traditional biotechnology involved for example, developing a new wheat variety with early ripening characteristics by cross-breeding different types of wheat until the desired characteristics and only those characteristics were present in the new wheat variety. The terms 'gene technology', 'genetic engineering' and 'genetic modification' mean the same thing, and refer to one type of modern biotechnology. Since the 1940s scientists have known that DNA — deoxyribonucleic acid — in the cells of all living things is like a blueprint that is passed from one generation to another. Genes are made of DNA. They contain coded instructions for proteins, which give living things their particular characteristics like hair and eye colour. All living organisms use DNA as the genetic code. Gene technology includes a range of techniques to copy genes and modify them so that they will work in a new host either to produce or reduce a product. It also involves transfer of the reconstructed gene to a new host. Using gene technology, scientists aim to introduce, enhance or delete particular characteristics of a living thing, depending on whether they are considered desirable or undesirable. As you go further into this website, you can read more about gene technology — its history, how it's done, its possible uses, the regulatory processes in place for gene technology, the debate surrounding gene technology and how this technology might affect you.

Biotech and Gene Technology

Gene technology is a term that refers to a whole range of techniques for genetic analysis that depend on the direct manipulation of DNA, the material substance of heredity. The use of gene technology in biomedicine, agriculture, food production and processing is an issue that has evoked strong public interest and concern. The debate has focused on the use of genetically modified organisms (GMOs) in food, their safety and potential impact on the environment. The public better accepts medical and industrial uses of gene technology. The majority of Australians are eager for more quality information about the technology and its applications.

Most people think of gene technology as adding genes from other life forms to plants and animals. While this is certainly part of it, the technology is broader than that, and has many more uses. Gene technology is developing at a rapid rate and it will continue to revolutionise basic biological research and development. It provides the potential to improve our health, create a safer and more secure food supply, generate greater prosperity and attain a more sustainable environment. Gene technology is already providing new ways of preventing, treating and curing human and animal diseases; it is helping farmers improve agricultural production with less impact on the environment; and in the near future, it will allow better food products to be available to consumers at reduced cost. The technology has given us vital new products like human insulin for diabetes, interferon and other drugs for treating certain cancers, and vaccines against diseases like hepatitis B. Millions of human lives are protected by gene technology every day. Genes are made from a chemical called DNA (deoxyribonucleic acid) that is common to all forms of life on Earth. Genes carry information for making all the proteins required by all organisms. These proteins determine, among other things, how the organism looks, how well its body uses food or fights infection, and sometimes even how it behaves. DNA is made up of four similar chemicals (called bases and abbreviated A,T, C, and G) that are repeated millions or billions of times throughout a genome. A genome is the entire DNA in an organism, including its genes. The human genome, for example, has 3 billion pairs of bases. The particular order of As, Ts, Cs, and Gs is extremely important. The order is called DNA sequence and variations in sequence underlie all of life's genetic diversity. DNA sequence is the database determining how the cell is to make proteins and what functions the proteins can perform. Because all organisms are related through similarities in DNA sequences, insights gained from non-human genomes often lead to new knowledge about human biology, and insights from the genes of a bacterium, worm or bee might contribute to new solutions for the wheat farmer. Gene technology consists of tools and techniques that scientists can use to study, identify or modify the genes of living organisms.

Techniques and Tools

Genetic modification

Over 20 years ago it was discovered that genes, and parts of genes, could be extracted from DNA using protein "scissors" then copied, or cloned. The gene could then be equipped with genetic "switches" to turn it up or down and inserted back into a living organism. This is the basis of genetic modification. It allows the sequences of genes to be determined. Transferring single genes between different plants and animals, turning existing genes up or down, or removing a gene from its original position and placing it in a new position in the same organism are all referred to as *genetic modification (GM)*. The terms genetic engineering (GE) or genetic manipulation also mean the same thing. Plants, animals or microbes which have a new gene inserted into them are called genetically modified organisms (GMOs) or transgenics. The modified gene belongs to the new host but its sequence is altered for example, on the basis of specific information gleaned from studies of other organisms.

DNA sequencing

DNA sequencing is the process of determining the exact order of the chemical building blocks (bases) that make up DNA. It is a laboratory procedure that involves first breaking down the DNA into short pieces, followed by separating the individual fragments using a technique called gel electrophoresis. A bar code pattern of the DNA pieces is produced which can then be read by computer. An enormous volume of information on the genetics of organisms is being generated using this technique and is providing computer experts with a great challenge in handling the data.

Human Genome Project

The Human Genome Project (HGP), which began in 1990, was a 13-year international project involving Australia. Its goal was to discover the approximate 50,000 genes that make up the human genome, make them accessible for further biological study and to determine the complete sequence of the 3 billion DNA subunits. Knowledge gained through this research will allow scientists and doctors of the future to prevent or treat most diseases at the level of genes. While it is now common to screen infants for inherited genetic diseases, further understanding of the human genome will allow doctors to determine a person's chances of developing particular diseases later in life. They will then be able to give advice or treatment to prevent, slow or cure those diseases.

Cloning Technology

A "clone" is a copy of a plant, animal, microcrganism or gene derived from a single common ancestor gene, cell or organism. Identical twins are natural clones from the one fertilised egg. A plant cutting is a clone of the original plant. Some confusion arises because the word *clone* is also applied to genes. A gene is said to be cloned when its sequence is multiplied many times in a common laboratory procedure. The cloning of a gene is a step in determining its sequence and initiating many types of experiments to understand its function and biology. The possibility of human cloning, raised when Scottish scientists at the Roslin Institute created the much-celebrated sheep "Dolly" in 1996, aroused worldwide interest and concern because of its scientific and ethical implications. Dolly was produced by injecting a nucleus from a mammary cell into an egg cell without a nucleus and raising an animal from that egg by treating it as a fertilised egg. Dolly was therefore a genetic clone of the animal from which the nucleus was taken. The important difference from identical twins is that the donor was a mature sheep and could even be made the mother of the clone. Cloning of animals and plants can be used to develop more efficient ways to produce superior breeds of animals. Cloning also enables genes to be added (such as those for human proteins) to produce animals that can generate hormones and other pharmaceuticals in milk, eggs or other products. The governments of many nations involved in gene technology have banned the cloning and genetic modification of human individuals. However, cloning technology may soon be available for human benefit to produce whole organs or special tissues from single cells for transplant to humans. It should be possible to use cells from the patient's own body and correct genetic defects before growing the tissue for transplant back into the donor. This is called therapeutic cloning.

Gene Marker Technology

DNA probes in diagnosis

Scientists can use gene technology to locate and analyse single genes in a chain of many thousands. They can make gene probes that recognise DNA sequences associated with genetic diseases. The diagnostic tests involve analysing minute tissue samples collected from adults, or taken from embryos. This information can assist parents and doctors to reduce the incidence of crippling genetic defects, and treat other disorders much better.

Forensic DNA fingerprinting

No two individuals have the same genetic makeup, a fact that is now being used by forensic scientists to identify individuals from cell samples left at a crime site. "Forensic DNA fingerprinting" involves taking a blood or tissue sample from a person and using one of a number of gene technologies to generate patterns of DNA fragments separated by size. Those patterns look something like a long bar code and are dependent on the exact sequences in the sample. A match between the genetic fingerprint from the crime sample and the suspect has been admissible evidence for a number of years.

DNA markers in breeding

For centuries plants and animals have been selectively cross-bred for desirable traits such as size, enhanced quality and pest resistance. Scientists can now use DNA information to identify desirable traits in organisms. These DNA markers are helping breeders to select superior strains of plants or types of animals for commercial production. The technique is similar to the forensic DNA fingerprint but where a particular DNA bar is known to be associated with a specific superior trait. By using these markers superior animals or plants can be identified easily and early and these individuals selected for further breeding to improve the herd or crop. CSIRO research on gene markers includes:

- markers in cattle for carcass weight, fat content, colour, meat tenderness and leanness. This information assists breeders to select superior types of animals.
- in fish farming, markers are helping breeders of super prawns select for desirable traits like size and colour, disease resistance and environmental suitability.
- in crop breeding, markers are being used to combine genes for disease resistance and quality characteristics, such as higher yield and bread making properties in wheat.

Most animals, plants and fish are still bred in the same way they have been for over 2000 years — but gene technology means that breeders can now pick the best parents with greater precision.

Transgenics

Because DNA is the same in all living things, researchers have found it possible to use genetic information derived from one organism in the modification of the genes of another. It is possible to modify the physical characteristics in precise ways. The term 'transgenics' often used to describe transferring genes between species, but in reality the process is a sharing of genetic knowledge for specific outcomes. In practice, the modified gene often has some of the gene sequences and signals from the host with some sequence information provided by the donor. Gene transfers typically involve the exact modification of one gene while leaving the other 50,000 genes unaffected. Over the last 25 years, researchers have learned how to make precise genetic changes by identifying and transferring individual genes to produce a desired characteristic. Scientists can insert new genes into an organism to add a particular trait not found in that organism. The new organism is then tested over several generations before researchers know whether the modified organism has all the intended benefits. Methods for inserting a gene into an animal or plant include:

- microinjecting into an animal egg using a fine tube
- using harmless viruses to carry genes into a host
- inserting DNA into plant cells on gold or tungsten particles
- using *Agrobacterium,* a bacterium that naturally transfers DNA into plants.

Genes are being introduced into plants to produce:

- resistance to insect attack or microbial damage, so reducing pesticide sprays and natural carcinogenic aflatoxins (mould toxins)
- tolerance to benign herbicides, so weeds can be controlled and erosion minimised through minimum tillage farming
- higher quality protein and enhanced vitamin or iron levels
- healthier vegetable oils and starches.

Genes are being inserted into animals to make them:

- produce more meat and less fat
- grow faster
- produce more meat or milk for the same amount of feed
- become more resistant to disease and parasites.

Because the techniques used to transfer genes sometimes have a low success rate, before a gene is transferred, another gene — called a "selectable marker gene" - is sometimes attached to the target gene to let researchers know whether the new gene is present or not. Scientists often use genes for antibiotic-resistance as selectable markers to show if plants have taken up a new gene. This practice allows only the genetically modified cells to grow in a special culture that contains the specific antibiotic. The antibiotics are used only in the laboratory and are of no medical significance. The antibiotic resistance genes employed are already widely distributed through the natural environment, such that their presence in the plant is of no environmental consequence. Nevertheless, because of public concerns about antibiotic-resistance genes, researchers have developed ways to remove these gene markers from plants before they are released commercially. New types of marker genes have also been developed, such as those permitting growth in the laboratory on unusual sugars.

Gene silencing

A gene that is producing undesirable characteristics in an organism can be turned down or switched off. One way this can be achieved is by inserting a second copy of the gene, or a fragment of the gene, back to front. Other ways include RNAi and "gene shears", which cause the gene message molecule to destroy itself. This cancels the effect of the undesirable gene. Gene silencing is being used to:

- protect plants from viruses
- alter the colour of ornamental flowers
- delay ripening of tomatoes and peaches so they reach the consumer in better condition
- remove allergens (proteins that cause allergies) from peanuts, soybeans and wheat
- stop undesirable browning in potatoes and raw sugar.

Gene therapy

Many human and animal diseases have now been traced to their host being born with faulty genes that produce defective proteins. These diseases include sickle cell anaemia, cystic fibrosis, the blood clotting disorder haemophilia and Down's syndrome. Modern medicine and surgery have made great strides in treating patients with genetic diseases. Scientists are now exploring the use of gene therapy to attack these diseases at their source — by correcting faulty genes. Gene therapy holds great promise for treating disease by replacing or changing a very small part of the overall genetic program of carefully selected cells, perhaps permanently, producing a cure. It aims to restore the healthy function of cells by replacing or correcting the defective gene. Gene therapy can be used to replace an abnormal gene with a normal one, to insert a missing gene, to switch off rogue genes that may cause cancers and to stop viruses multiplying within cells. The modified cells and genes are not passed onto children. Certain technical obstacles are yet to be overcome if gene therapy is to deliver the expected benefits to treat human disease. Among the obstacles is the lack of a delivery system, or vector, that can safely and efficiently shuttle beneficial genes into the cells of patients — and ensure they work. Even the most advanced cell therapy techniques are still at the experimental stage in clinical trials. Further research is needed to develop safe, reliable gene therapy techniques.

Prerequisite Course and Institutions

Bio-Organic Chemistry

This course provides a foundation in structural organic chemistry, acid base chemistry, chemical thermodynamics, and reaction mechanisms. Subjects include Lewis structures, atomic and hybridized orbitals, stereochemistry, inter- and intramolecular forces of attraction, neucleophilic reaction mechanisms, functional groups, and the organic chemistry of biological molecules. Please note that this course does not count toward requirements for the master's degree in biotechnology.

Bioscience for Regulatory Affairs

This course examines the fundamental underlying scientific concepts utilized in the creation and development of biomedical products. Topics to be covered include the structure and function of biomolecules such as proteins, enzymes, carbohydrates, lipids, and DNA, as well as the structure and function of cellular components such as membranes, vesicles, organelles, and the cytoskeleton. In addition, students will examine the complexities of metabolism, DNA replication, transcription, translation, signal transduction mechanisms, apoptosis, the cell cycle, and cancer.

Core Science Courses

Students must complete four core courses before they are allowed to enroll in most of the science electives. A student who has mastered some or all of the material covered in the core courses in previous academic work may request that one or more of these courses be waived. After reviewing the student's request, the program committee may approve a waiver. The student then is required to replace the waived core course(s) with science electives. Electives should be chosen in consultation with the student's adviser and should accommodate individual career goals.

Biochemistry

This course explores the roles of essential biological molecules focusing on protein chemistry, while covering lipids and carbohydrates. It provides a systematic and methodical application of general and organic chemistry principles. Students examine the structure of proteins, their function, their binding to other molecules and the methodologies for the purification and characterization of proteins. Enzymes and their kinetics and mechanisms are covered in detail. Metabolic pathways are examined from thermodynamic and regulatory perspectives. This course provides the linkage between the inanimate world of chemistry and the living world of biology.

Molecular Biology

This course provides a comprehensive overview of the key concepts in molecular biology. Topics to be covered include nucleic acid structure and function, DNA replication, transcription, translation, chromosome structure and remodeling, and regulation of gene expression in prokaryotes and eukaryotes. Extended topics to be covered include methods in recombinant DNA technology, microarrays and microRNA.

Advanced Cell Biology I

This course covers cell organization and subcellular structure. Students examine the evolution of the cell, chromosome and plasma membrane structures and behaviors, mechanics of cell division, sites of macromolecular synthesis and processing, transport across cell membranes, cell dynamics, organelle biogenesis, and cell specialization. Students also are introduced to the experimental techniques used in cell biology to study cell growth, manipulation, and evaluation.

Advanced Cell Biology II

This course is a continuation of 410.603 (Advanced Cell Biology I) and further explores cell organization and subcellular structure. Students examine cell-to-cell signaling that involves hormones and receptors, signal transduction pathways, second messenger molecules, cell adhesion, extracellular matrix, cell cycle, programmed cell death, methylation of DNA and modification of chromatic structure, and mechanisms of the cell. The involvement of abnormalities in signal transduction pathways to oncogenesis and other disease states will be stressed.

Gene Organization and Expression

Students use genetic analysis and molecular biology techniques to investigate chromosome organization, chromatin structure, functional genomics, and mechanisms of differential gene expression. Other topics include DNA methylation, silencers, enhancers, genomic imprinting, and microarray analysis.

Vaccinology

This course will cover the biological development, immunologic concepts, and methods for vaccine delivery. Specific topics include new technologies for vaccine development such as DNA vaccines, recombinant mucosal vaccines, dendritic cells for antigen delivery, novel adjuvants, and methods to increase vaccine stability. Delivery systems for vaccines, both time tested and new methodologies such as lipid-based systems, needle-free injection systems, and novel methods such as the use of genetically modified foods, will be discussed. The underlying biological role of the innate and adaptive immune systems will be explored in light of new types of vaccines and delivery systems. Finally the process of bringing vaccines to market will be covered including government oversight and licensure.

Human Molecular Genetics

In this course students learn to use the tools of modern genomics to elucidate phenotypic variation within populations. The course uses human disease (from simple Mendelian disorders to common complex disorders) to exemplify the types of studies and tools that can be used to characterize cellular pathophysiology as well as to provide genetic diagnostics and therapies. Students become facile with linkage analysis, cancer genetics, microarray analysis (oligo and DNA arrays), gene therapy, SNP studies, imprinting, dis-equilibrium mapping, and ethical dilemmas associated with the Human Genome Project.

Microbiology

This course is an overview of microorganisms important in clinical diseases and biotechnology. Students are introduced to the general concepts concerning the morphology, genetics, and reproduction of these microbial agents. Lectures focus on individual organisms with emphasis on infectious diseases, biotechnology applications, molecular and biochemical characteristics, and molecular and serological identification methods. Students will also discuss the impact biotechnology, and particularly genomics, will have on the development of antibiotics and vaccines as treatment and preventive measures.

Principles of Immunology

This course covers molecular and cellular immunology, including antigen and antibody structure and function, effector mechanisms, complement, major histocompatibility complexes, B- and T-cell receptors, antibody formation and immunity, cytotoxic responses, and regulation of the immune response. Students are also introduced to the applied aspects of immunology, which include immunoassay design and flow cytometry. Special topics include immunomodulation, immunosuppression, immunotherapy, autoimmunity, and vaccination.

Pathogenic Bacteriology

Lecture and discussion augmented by guided readings on pathogenic bacteria, with special attention to microorganisms that cause human disease. The course is designed to impart to the student an appreciation as well as knowledge of the history, epidemiology, cultivation, morphology, serology, biochemistry, and clinical description of the major disease-producing bacteria. Discussion of therapeutic considerations and vaccination will also be included in this course.

Virology

This course covers the advanced study of viruses with regard to the basic, biochemical, molecular, epidemiological, clinical, and biotechnological aspects of animal viruses primarily and bacteriophage, plant viruses, viroids, prions, and unconventional agents secondarily. Specific areas of virology, including viral structure and assembly, viral replication, viral recombination and evolution, virus-host interactions, viral transformation, gene therapy, antiviral drugs, and vaccines, are presented. The major animal virus families are discussed individually with respect to classification, genomic structure, virion structure, virus cycle, pathogenesis, clinical features, epidemiology, immunity, and control. The viral vectors and their applications in biotechnology are discussed.

Marine Biotechnology

This course covers the application of molecular techniques

to study the marine environment and obtain useful products from marine systems. Students examine recent progress in discovery of drugs and enzymes from marine microbes and macroorganisms, biodiversity, bioremediation, molecular approaches in aquaculture, the role of marine microbes in global carbon cycling, and genomics of marine organisms.

Parasitology

This course concentrates on the developmental biology, natural history, and cell and molecular biology of the major eukaryotic parasites that cause significant human morbidity and mortality in the world. The fundamental mechanisms in host-parasite relationships, diagnosis, pathogenesis, epidemiology and control strategies will be emphasized. The major focus will be *Plasmodium spp*, the causative agent of malaria (number 3 among the top 10 single disease killers in the world). Other parasites that will be studied include *Trypanosoma* (African trypanosomiasis, Chagas disease), *Schistosoma* (Schistosomiasis), *Leishmania* (Leishmaniasis) *Onchocerca* (Onchocerciasis), other blood and intestinal helminthes, tapeworms and *Pneumocystis jiroveci*. Current research in parasitology will be discussed.

Molecular Evolution and Phylogenetics

This course covers the principles of molecular evolution and phylogenetics. Topics include patterns and analyses of DNA polymorphism, genetic evolutionary trees, molecular clocks, the evolution of multigene families, gene duplication and shuffling, transposition and horizontal gene transfer, gene number and genome size, organellar and nuclear genetic markers, genetic mutation and selection, genes in populations, viral evolution, human evolution, and the theoretical background for molecular phylogenetics. Examples of each concept will be drawn from the scientific literature in epidemiology and human or animal genetics.

Agricultural Biotechnology

In this course, students are introduced to the application of recombinant DNA technology to agriculture. Studied are methods for the introduction of foreign DNA into plant and animal cells and generation of stably transformed plants and animals. Students consider specific examples of the use of transgenic plants and animals in biotechnology, which can provide protection against insects, diseases, and tolerance to specific herbicides. They also investigate how recombinant growth hormones can result in leaner meat, greater milk yield, better feed utilization, and how transgenic plants and animals can serve as bioreactors for the production of medicinals or protein pharmaceuticals. Because recombinant agricultural products are released into the environment or consumed as foods, students also need to become familiar with environmental safety issues.

Molecular Basis of Pharmacology

This course begins by reviewing receptor binding and enzyme kinetics. Various cellular receptors and their physiology are discussed as well as the pharmacological agents used to define and affect the receptor's function. Students study the pharmacology of cell surface receptors and intracellular receptors. Also considered are the drugs that affect enzymes.

Molecular and Cellular Physiology

Students in this course gain an understanding of how coordinated regulation of bodily function occurs at the molecular and cellular levels of organization. The focus is on neurons, muscles, and hormones. Specific areas covered for excitable tissue include bioelectric properties of excitable membranes, Hodgkin-Huxley ion currents, voltage-gated ion channels and their structures, synaptic transmission, excitation-contraction coupling, and contractile properties of skeletal, cardiac, and smooth muscle cells. The biotechnological connection is the pharmacological interventions to modulate functioning of excitable tissues. For endocrine physiology, there is an overview of hypothalamic, pituitary, reproductive, and other hormones. The class uses leptin and obesity as a model hormone and pathology, respectively, and examines in detail its action as a putative fat busting hormone.

System and Integrative Physiology

This course is the second half of the Physiology sequence and involves the study of organ systems and how they are regulated by the central nervous and endocrine systems. Students will learn the structure and function of the cardiovascular, respiratory, digestive, renal, and reproductive systems, as well as their pathophysiology during disease processes. We will also study metabolic physiology in the context of exercise and diet. The biotechnological connection will be how the drug interventions modulate functioning of many of these systems.

Industrial Microbiology

This course covers the principles of various processes associated with the production and recovery of different bioproducts derived from prokaryotes and eukaryotes. Topics include the classification of microorganisms, media development, instrumentation, fermentation principles, mammalian and insect cell propagation, product recovery, protein purification, and the principles of current good manufacturing practices (cGMP). Emphasis is on large-scale production methods and production of recombinant proteins for diagnostic and clinical applications.

Molecular Development

This course covers the molecular and cellular bases of development in a variety of experimental organisms with special emphasis on mammalian and human models. From the formation of germ cells, fertilization, and early embryonic development to the final formation of organs and tissues, developmental processes are considered in the context of biotechnological applications. Application possibilities include creation of transgenic animals and drug design to combat specific types of cancer. The molecular mechanisms of developmental processes and the identification of targets for therapeutic purposes are central themes.

Product Development: From Research to Manufacturing

This course provides an extensive overview of a process for development of a pharmaceutical by a biotechnology company. The course emphasizes the importance of intellectual property, the basic sciences underpinning the development of a product, and the importance of the interaction between a biotechnology company and the Food and Drug Administration (FDA). Students learn to appreciate the importance of quality control and assurance, good manufacturing practices, preclinical and clinical testing, and the lengthy regulatory processes which govern the development, manufacture, and eventual sale of biotechnological products. Hands-on solving of practical problems and guest lecturers who are experts in the field familiarize students with the intricacies of the process.

Neurobiology

This course provides a framework for understanding the molecular physiology of neuronal structure, signaling, and circuitry and how this cellular design is ultimately integrated to achieve higher cognitive functions such as perception, control of movement, learning, and memory. The course introduces the students to various current neuroscience topics, including but not limited to membrane physiology and electrical excitability of neurons, neurotransmitters and synaptic transmission, signaling at the neuromuscular junction, cellular and higher-order aspects of perception and motor control, molecular mechanisms of neural development, and the molecular basis of learning and memory. This course places particular emphasis on the genetic and molecular bases of a wide variety of neurological and neurodegenerative diseases such as multiple sclerosis, amyotrophic lateral sclerosis, Parkinson's, and Alzheimer's.

Genes and Disease

Because of recent advances, powerful diagnostic tests now detect genetic diseases, and there is promise of gene replacement therapy. In this course students cover general genetic principles, DNA tools for genetic analysis, cytogenetics, gene mapping, the molecular basis of genetic diseases, animal models, immunogenetics, genetics of development, genetics of cancer, and treatment of genetic diseases. Molecular methods of analysis are emphasized.

Gene Therapy

Students are introduced to gene transfer, its technical evolution, and its testing through clinical studies. Gene therapy holds promise for both genetic diseases and acquired diseases such as cancer and AIDS. The health, safety, and ethical issues surrounding gene therapy are discussed, together with the review and oversight systems established to regulate this therapy. Students also consider how industry is developing these techniques, both in new start-up companies as well as in established biotechnology and pharmaceutical companies. An overview of proprietary and patent issues in gene therapy is part of the course.

Infectious Diseases

This course focuses on infectious diseases of mankind, presented in a system-by-system format. Basic principles of host defense and microbial virulence will be discussed. Practical up-to-date information on the clinical presentation, symptoms, physical findings, laboratory diagnosis, treatment, and prevention of the general array of diseases caused by bacteria and viruses will be presented. The use of antibiotics, prophylactic agents, and vaccines along with selected aspects of pathogenesis and epidemiology will be covered. More cursory coverage will be given to the fungal and parasitic agents of human disease. The student will develop a broad understanding of the many different kinds of infectious processes that our bodies are subjected to on an ongoing basis.

Emerging Infectious Diseases

This course focuses on emerging infectious diseases from many different perspectives. The maladies addressed range from diseases that have reappeared in altered genetic forms, such as the influenza virus and the West Nile virus, to the lethal hemorrhagic fever caused by the Ebola virus. Also discussed is the threat of recombinant and ancient infectious agents such as Bacillus anthracis, causative agent of anthrax, which can be used in biological warfare weapons. Opinions from noted scientists and leaders concerning emerging diseases and the prospects for battling them successfully provide scientific and social perspective.

Computers in Molecular Biology

Retrieval and analysis of electronic information are essential in today's research environment. This course explores the theory and practice of biological database searching and analysis. In particular, students are introduced to integrated systems where a variety of data sources are connected through World Wide web access. Information retrieval as well as interpretation are discussed, and many practical examples in a computer laboratory setting enable students to improve their data mining skills. Methods included in the course are searching the biomedical literature, sequence homology searching and multiple alignment, protein sequence motif analysis, and several genome analytical methods. Classes are held in a computer laboratory. Acquaintance with computers is required.

Practical Computer Concepts for Bioinformatics

This course introduces students with a background in the life sciences to the basic computing concepts of the UNIX operating system, relational databases, structured programming, object-oriented programming, and the Internet. Included is an introduction to SQL and the Perl scripting language. The course emphasizes relevance to molecular biology and bioinformatics. It is intended for students with no computer programming background, but with a solid knowledge of molecular biology.

Bioinformatics: Tools for Genome Analysis

Several large-scale DNA sequencing efforts have resulted in megabase amounts of DNA sequences being deposited in public databases. As such, the sequences are of less use than those sequences that are fully annotated. To assign annotations such as exon boundaries, repeat regions, and other biologically relevant information accurately in the feature tables of these sequences requires a significant amount of human intervention. This course instructs students on computer analytical methods for gene identification, promoter analysis, and introductory gene expression ananlysis using software methods. Additionally, students are introduced to comparative genomics and proteomic analysis methods. Students will become proficient in annotating large genomic DNA sequences. Students complete two large sequence analysis projects during the course.

Biology of HIV and AIDS

This course includes an overview of the biology and life cycle of the immunodeficiency virus, including the simian viruses (SIVs). Specific areas of HIV immunolopathogenisis are emphasized, to include HIV diagnosis, HIV-induced immune dysfunction, and therapeutic breakthroughs in the treatment of HIV-1 disease. Students become familiarized with current methods in biotechnology that have advanced our understanding of the biology of retroviruses. Special topics include international genetic variation (subtypes and clades),. HIV vaccine development, and global economic impact.

Cancer Biology

This course provides students with knowledge of the fundamental principles of the molecular and cellular biology of cancer cells. Lectures and demonstrations explain the role of growth factors, oncogenes, tumor suppressor genes, angiogenesis, and signal transduction mechanisms in tumor formation. Discussion of aspects of cancer epidemiology, prevention, and principles of drug action in cancer management is part of the course.

Protein Bioinformatics

Because the gap between the number of protein sequences and the number of protein crystal structures continues to expand, protein structural predictions are increasingly more important. This course provides a working knowledge of various computer-based tools available for predicting the structure and function of proteins. Topics include protein database searching, protein physicochemical properties, secondary structure prediction, and statistical verification. Also covered are graphic visualization of the different types of three-dimensional folds and predicting 3-D structures by homology. Computer laboratories complement material presented in lectures.

Phylogenetics and Comparative Genomics

This course will provide a practical, hands-on introduction to the study of phylogenetics and comparative genomics. Theoretical background on molecular evolution will be provided only as needed to inform the comparative analysis of genomic data. The emphasis of the course will be placed squarely on the understanding and use of a variety of computational tools designed to extract meaningful biological information from molecular sequences. Lectures will provide information on the conceptual essence of the algorithms that underlie various sequence analysis tools and the rationale behind their use. Only programs that are freely available, as either downloadable executables or as web servers, will be used in this course. Students will be encouraged to use the programs and approaches introduced in the course to address questions relevant to their own work.

Clinical and Molecular Diagnostics

This course covers basic concepts and practical applications of modern laboratory diagnostic techniques. Topics include the principles of testing methodology, quality assurance, and the application of molecular methods to the clinical and research laboratory. The test methods to be covered include nucleic acid-based methods such as hybridization, amplification, and sequencing; non-nucleic acid methods such as HPLC, GLC, and protein analysis; and technologies such as PFGE, ribotyping, RFLP, and serological testing methodologies. In addition to the test procedures, students are exposed to aspects of statistics, quality control, regulatory issues, and applications of these methods to the diagnosis and prognosis of human disease.

Biostatistics

This course introduces statistical concepts and analytical methods as applied to data encountered in biotechnology and biomedical sciences. It emphasizes the basic concepts of experimental design, quantitative analysis of data, and statistical inferences. Topics include probability theory and distributions; population parameters and their sample estimates; descriptive statistics for central tendency and dispersion; hypothesis testing and confidence intervals for means, variances, and proportions; the chi-square statistic; categorical data analysis; linear correlation and regression model; analysis of variance; and nonparametric methods. The course provides students a foundation to evaluate information critically to support research objectives and product claims and a better understanding of statistical design of experimental trials for biological products/devices.

Clinical Trial Design and Conduct

Through a case-study approach, this course will cover the basic design issues of clinical trials. The design of a specific trial will be studied to illustrate the major design issues, such as overall design, endpoint variables, control group selection, eligibility criteria, and potential ethical issues such as the use of placebo controls. The course also covers the planned analysis of these studies, including approaches that are central to clinical trials, such as stratified analysis, adjustment factors, and "intention-to-treat" analyses. Also to be covered are the operational aspects of clinical trial conduct (Good Clinical Practices). Group projects include a full protocol for a clinical trial of a fictitious new drug or biologic, plus the accompanying informed consent document. The semester will conclude with a mock IRB meeting to review one of the student projects.

Clinical Development of Drugs and Biologics

This course introduces students to the planning and work required to develop potential new drugs and biologics efficiently. Students gain a thorough appreciation of FDA and ICH regulations and guidelines. Because the course emphasizes the importance of planning before the execution of any of the necessary steps, lectures use a "backwards" approach, discussing the final analysis and report before developing protocols. Topics also include an overview of preclinical investigations; NDA/BLA format and content; clinical development plans; product and assay development; the IND ; and trial design, implementation, and management.

Tissue Engineering

Tissue engineering is a highly multidisciplinary field that involves cell biology, chemistry, materials science, engineering, and medicine. This course will be a survey that introduces students to the field from scientific, clinical, manufacturing, and regulatory perspectives. Roughly the first half of the course will be devoted to background material, and the second half will focus on applications. Readings will be drawn from books and journals.

Transplantation Science

This course is designed to provide exposure and insight into transplantation science. The course emphasizes the current status of organ donation, procurement programs, and histocompatibility testing. In addition, immunological consequences of organ or tissue transplantation and subsequent immuno-suppressive regimens are examined. Special attention is given to promising and innovative technologies such as tissue engineering, cardiac regeneration, stem cell technology, and xenotransplantation. Students investigate the biological and ethical issues associated with these new technologies.

Radiation Biology (Updated)

This course will review types of ionizing radiation and their differences, physical and chemical interactions of radiation with key biological molecules, effects on living matter beginning with molecular and cellular interactions and proceeding to tissue, organ, and organism levels, emphasizing the human system. Radiation's beneficial effects in cancer therapy and medicine as well as detrimental and carcinogenic effects will be discussed. Specific units will consider food irradiation, nuclear power plant accidents, radiation terrorism, everyday sources of exposure to the US population, and other practical situations involving radiation.

Methods in Proteomics

This course covers the analytical methods used to separate and characterize pharmaceutical compounds (predominantly proteins) derived through biotechnology. While emphasis is placed on the general principles and applicability of the methods, current protocols are discussed, and problem sets representing realistic developmental challenges are assigned. Topics include chromatography (HPLC, SEC, IEC), electrophoretic techniques (2-D gel electrophoresis), spectroscopic methods (UV/Vis, fluorescence, CD), analytical ultracentrifugation, microarrays, mass spectroscopy, amino acid analysis, sequencing, and methods to measure protein-protein interactions.

Epidemiology: Diseases in Populations

Epidemiology is the study of the patterns and determinants of disease in populations. It constitutes a basic science for public health and biomedical sciences and its influence can be felt daily through the presentation of data by government, academic, and industry sources. The goal of this course is to present an introduction to epidemiological methods and inferences to biotechnology professionals with little prior experience in public health. Issues in epidemiological inference and the assessment of causal relationships from epidemiological studies will be discussed, introducing the issues of bias and confounding. Throughout the course, emphasis will be on the practical use of epidemiology and lectures will be complemented by case studies and published literature. Examples will be drawn from contemporaneous issues in chronic and infectious diseases. At the conclusion of the course, students should have a greater appreciation for the role of the epidemiologic method and be able to evaluate a basic epidemiologic study, including how the study goals and research questions relate to the design, measures, and inferences.

Current Topics in Molecular and Cellular Biology

This course is a literature-driven exploration of current topics and methodologies employed in cell biology research. By closely examining both fundamental and innovative experimental approaches, as well as cutting-edge technologies, we will explore a broad range of cell biology topics, many of which most students have previously encountered, at least at some level, in the core courses. The particular subjects and technologies discussed may vary widely from semester to semester; topics may include but are not limited to expression profiling (microarray analysis),"knock-down" with RNAi, the use of transgenic and knock-out mice, proteomics and mass spectrometry, microscopy applications (epifluorescence, confocal, and/or EM), characterizing protein/protein interactions, and detection methods for the movement of small molecules and ions. Students enrolling in this course are expected to already have some experience in critical reading and evaluation of the primary scientific literature.

Genomic Sequencing and Analysis

The completion of the human genome sequence is just the latest achievement in genome sequencing. Armed with the complete genome sequence, scientists need to identify the genes encoded within, to assign functions to the genes, and to put these into functional and metabolic pathways. This course will provide an overview of the laboratory and computational techniques beginning with genome sequencing and annotation, extending to bioinformatics analysis and comparative genomics, and including functional genomics.

Theory and Applications of Immunoassays

Antibodies are useful as molecular tools in a variety of applications in biotechnology. They can be produced quickly, inexpensively, and in very large quantities. Students will examine how antibodies can be configured as a measurement tool called an immunoassay. Students design immunoassays for their own laboratory applications such as radio- and enzyme-immunoassays, and competitive and immunometric immunoassays. They learn how to modify antibodies chemically for conjugation of reporter molecules; and they become aware that various immunoassay performance issues can affect results. Students also investigate immunoassay formats such as configuration of antibodies as biosensors and surface plasmon resonance signaling.

Microarrays and Analysis

This course will focus on the analysis and visualization of microarray data. The general aim is to introduce students to the various techniques and issues involved with analyzing multivariate expression data. Additionally, students will visualize the results in modern statistical scripting software. Topics include detecting and attributing sources of data variability, determining differentially expressed genes with relevant statistical tests, and controlling for false positive discovery (multiple test corrections, permutations, etc.). An introduction to linear and non-linear dimension reduction methods (PCA, PLS, isometric feature mapping, etc.) and an introduction to common pattern recognition (clustering), classification, and discrimination techniques will be included. Assignments and concepts will make use of publicly available Affymetrix and cDNA microarray data sets. Examples will mostly be demonstrated in S-plus and R (publicly available) code, with some in SAS. Free demo software tools such as Minitab, Spotfire, TreeView, Expression Profiler, and web UIs will also be utilized.

Biothreat Response & Microbial Forensics (Updated)

This course introduces the methods and techniques used for biothreat detection, identification, and medical intervention. Methods to protect the health and safety of responders, the proper procedures for threat containment, decontamination, removal, and establishment of a chain of custody for evidentiary materials are all discussed.

Students will learn rapid methods for biothreat detection (using various biodetection platforms) and conventional methods such as culture, PCR, DNA fingerprinting, phage typing, microarrays, and proteomics. Procedures for working in Biosafety Levels 3 and 4 environments are presented, as well as the immunological responses to biothreat exposure, and the preventive and therapeutic options available.

Science, Medicine, and Policy in Biodefense

This course provides a comprehensive introduction to the Concentration in Biodefense. Biological warfare is introduced in its historical context, followed by the properties of the most important biological threat agents, their medical consequences and treatment, diagnostics and forensics. Relevant international and domestic policy issues are explored, along with defense strategies and the nature of existing dangers to national security. Students should leave the class with a deep understanding of biological warfare and terror agents, the consequences of their potential use, and the available means of protection.

Microbial Pathogens and the Impact on National Security

This course provides an overview of the basic microbiology, pathology, detection methods, and medical treatments required in response to a biothreat attack. Methods to protect the health and safety of responders and the general public are presented, and the procedures for biothreat weaponization and delivery, collection and storage, containment and decontamination, and risk assessment are discussed. The potential biothreat agents of bacterial, viral, and agricultural significance are presented, such as, Bacillus anthracis (anthrax), Yersinia pestis (plague), variola virus (smallpox virus), hemorrhagic fever viruses (Ebola virus, dengue virus), and microbiological toxins (botulinum, superantigens). *Non-science elective for Biotechnology program.*

Applied Molecular Biology

This course will cover applied concepts and research techniques in molecular biology. It is designed for students with a good basic knowledge of molecular biology who want to study more advanced concepts and how they may be applied in biotechnology. Topics for discussion include; applied PCR techniques, vector construction and modification, affinity chromatography, two-hybrid screening, iRNA, site-directed-mutagenesis /protein engineering, microarrays, prokaryotic and eukaryotic expression systems.

Bioassay Development

This course will cover methodological approaches to bioassay development for high throughput screening. Both cell-based (cytotoxicity; cytoprotection, high content imaging, and reporter systems) and cell-free assay systems (enzyme,

FRET, time resolved fluorescence, quenching assays, and immunological assays) will be included with discussion of the potential promise and pitfalls associated with each assay system. Various assay formats, visualization techniques, and current developments in assay technology will be discussed. Project management techniques will be utilized to aid in the process of assay development.

Microfluidics and Biosensors

Microfluidics (Lab-on-a-chip technology) is the miniaturization of laboratory operations for micro-scale chemistry, high-throughput drug screening, environmental sensors, biothreat detectors, forensics, clinical diagnostics and proteomics. This course will cover microfluidic implementations of bioassay development such as sample dilution, cell lysis, chromatography, solid-phase extraction, electrophoresis, nucleic acid amplification and sequencing, analyte detection, single-cell analysis, microarray design & mass spectrometry sample preparation. The materials, design, fabrication, and testing of microfluidic chips and biosensors will be discussed, with emphasis on the applications of this technology to detect microbial pathogens and cancer markers. In addition, the course will include case studies from the literature to introduce students to intellectual property issues related to microfluidic technology.

Bioperl

This course builds on the Perl concepts taught in 410.634 Practical Computer Concepts for Bioinformatics. Perl has emerged as the language of choice for the manipulation of bioinformatics data. Bioperl, a set of object-oriented modules that implements common bioinformatics tasks, has been developed to aid biologists in sequence analysis. The course will include an overview of the principal features of Bioperl and give students extensive opportunity to use Perl and the tools of Bioperl to solve problems in molecular biology sequence analysis.

Molecular Targets and Cancer

This course will investigate potential molecular targets in cancer including receptor tyrosine kinases, G-Protein coupled receptors, the TGF beta signaling pathway, cell cycle check points, kinases and phosphatases, chemokine and chemokine receptors, nuclear receptors, suppressor proteins, metastasis and angiogenesis targets, integrins, and matrix metalloproteinases. Discussion will also include topics on what defines a molecular target and the methods by which they are evaluated.

Chemical Libraries and Diversity

Chemical diversity and "pharmacological space" will be studied with an emphasis on disciplines related to drug discovery. Medicinal chemistry, natural product chemistry, focused synthetic libraries, and combinatorial chemistry will be covered. Lipinski's rules for drug-like molecules will be discussed in detail, as well as methods for chemical analysis, in silico drug design, molecular modeling, and compound storage and handling. In addition, techniques used for assessing and harnessing chemical diversity for drug discovery will be discussed.

Stem Cell Biology

This course will involve discussion and debate on current topics concerning stem cell biology and the use of stem cells in biotechnology and therapeutics. Topics will include review and discussion of developmental & cell biology, stem cell characteristics, stem cell preparation & therapeutic uses, tissue engineering, global regulatory and ethical issues, and commercialization of stem cell therapy. Current peer-reviewed literature and guest experts in the field will provide up to date information for discussion.

Comparative Microbial Genomics: From Sequence to Significance

Hundreds of Bacterial and Archaeal genomes have been completely sequenced and thousands more will follow in near future. In this course we will learn how to make sense of this vast sea of information in order to understand the diversity of microbial life on earth: transforming DNA data into knowledge about the metabolism, biological niche and lifestyle of these organisms. The use and development of bioinformatic platforms for the sensible comparison of genetic function and context are essential for work in modern microbiology. Topics covered will include methods for sequencing, gene finding, functional prediction, metabolic pathway and biological system reconstruction, phylogenomics, ontologies, and high-throughtput functional genomics. Particular attention will be paid to publically available bioinformatics resources and their proper use. Examples will be drawn from microbes of importance to human health, industry, ecology, agriculture and biodefense. Lectures and discussions are integrated with computer exercises where appropriate.

Independent Research in Biotechnology

Students in the biotechnology program have the opportunity to enroll in an independent research course. This elective course is an option after a student has completed at least eight graduate-level courses and has compiled a strong academic record. Prior to proposing a project, interested students must have identified a research topic and a mentor who is familiar with their prospective inquiry and who is willing to provide guidance and oversee the project. The research project must be independent of current work-related responsibilities as determined by the project mentor. The mentor may be a faculty member teaching in the biotechnology program, a supervisor from the student's place of work, or any expert with appropriate credentials. Students are required to submit a formal proposal for review and approval by the biotechnology program committee. The proposal must be received by the Advanced Academic Programs office no later than one month prior to the beginning of the term in which the student wants to enroll in the course. Students must meet with a member of the program committee periodically for discussion of the project's progress, and a written document must be completed and approved by the program committee and project mentor for the student to receive graduate credit. Additional guidelines can be obtained from the AAP administrative office. Download the Guidelines for Proposal. Sample IRPs. *Prerequisites : All four core courses and four elective courses.*

Laboratory Elective Courses

All the following electives are wet lab courses. Students must bring lab coat and safety glasses to all meetings of the course.

Cell Culture Techniques

This laboratory course illustrates the use of basic cell culture techniques used in bioscience research and commercial applications. Students are introduced to cell cultivation methods, including proper use of a biological safety cabinet, sterile technique, cell enumeration and media preparation, primary cultures, cultivation of cell lines, detection of contamination, cryopreservation, transfection, and an introduction to bioassays. This course is designed for students with no prior knowledge or with limited knowledge of cell culture methods.

Recombinant DNA Laboratory

This laboratory course introduces students to methods for manipulating and analyzing nucleic acids. Students gain extensive hands-on experience with plasmid purification, restriction mapping, ligations, bacterial transformations, gel electrophoresis, as well as applications of the polymerase chain reaction. This course is not recommended for students with substantial experience in these methodologies.

Recombinant Protein Expression, Production, and Analysis

This laboratory course introduces students to the construction, production, processing, and analysis of recombinant proteins from prokaryotic and eukaryotic sources. Concepts include the design, construction, and delivery of recombinant expression clones, expression of recombinant genes in host cells, protein purification, and protein analysis. Laboratory exercises use current techniques and approaches for the cloning, expression, production, purification, and analysis of recombinant proteins in bacteria and mammalian cells.

Biodefense Laboratory Methods

This laboratory course introduces students to the methods and techniques used for biothreat detection, surveillance and identification. Using bio-simulants and demonstrations, various biodetection platforms will be discussed and presented, such as point-of-detection devices and methods, laboratory based screening and identification technologies (culture, RT-PCR, ECL, immunoassays, biosensors), and high-throughput environmental surveillance methods. Statistical tools for determining diagnostic sensitivity & specificity, and assay validity will be discussed. Laboratory practices and procedures for working in simulated Biosafety Level 3 and 4 environments will be practiced. Students will be introduced to the current bioinformatics genomic and proteomic databases used for category 3 microbial identification and characterization.

Advanced Recombinant DNA Laboratory

This course is a continuation of 410.656 Recombinant DNA Laboratory, intended for those who completed the introductory course or who have extensive laboratory experience. This second course consists of a series of integrated laboratory exercises designed to give students hands-on experience with a variety of advanced recombinant DNA techniques. Exercises include cloning and analysis of an exon from a gene; creation of mutations within structural genes through the polymerase chain reaction (or transposon mutagenesis); and examination of mutations. Students isolate total RNA and quantitate a specific message using RT-PCR, examine gene expression using microarrays, and utilize bioinformatics tools for DNA analysis.

Immunological Techniques in Biotechnology

This laboratory course introduces students to methods for analyzing the immune system. Participants gain experience with various immunologic techniques used in research and biotechnology laboratories such as immunoassays, immunofluorescence, western blot analysis, SDS-PAGE, antibody purification (protein A), and cytokine assays. Additional topics for discussion include hybridoma technology, phage antibody libraries, therapeutic monoclonal antibodies, and flow cytometry.

High Throughput Screening and Automation Laboratory

This course will utilize hands-on instruction in automated bioassay systems for high throughput screening (HTS) as an entry point to covering pertinent aspects of HTS, such as data manipulation, storage, and analysis; liquid handling robotics; microtiter plate washing, manipulation and bar coding; HTS assay detectors; and automated devices for assay setup, validation and visualization. Cost considerations, HTS amenable assay systems, and miniaturization and scale-up will also be discussed.

Enterprise and Regulatory Affairs Elective Courses

Bioethics

Students in this course analyze and discuss traditional philosophical theories regarding the nature of the moral good. They then apply these theories to critical issues and selected cases involving experiments with human subjects, organ transplantation, in vitro fertilization, the use of animals in research, the collection and publication of research data, peer review, conflicts of interest, and other topics of current concern.

Economic Aspects of Biotechnology

This course examines the economic, business, and financial forces that define and shape the biotechnology industry, with special focus on early stage and growing companies. A major focus is on decisionmaking under uncertainty, which includes how to interpret and screen various input types in the decision process and how to balance the empirical and intuitive approaches to analyzing problems arising from the nature of the biotechnology business. The course will permit the student to prepare an analytic approach to complex problems, identify and assess risks and payoffs, evaluate risks and payoffs in monetary and non-monetary terms, develop a structure for strategy and tactics, and understand how and why people make the decisions they make. Topics include defining and distinguishing the biotechnology industry, competitive forces and impact on strategy, regulation of genetic products, planning under uncertainty, system thinking and system failure, the economic environment, estimating costs and benefits, strategic components, marketing and sales, research and development, and valuing a company

Managing Biotechnology Professionals

The roles of managers and leaders within biotechnology

companies undergo constant change. Biotechnology managers must engage in new and innovative problem-solving strategies; lead a diverse and global workforce; develop partnerships with other businesses, customers, and competitors; manage horizontally and across teams; and utilize technology as a competitive advantage. The student is able to address current challenges in his/her own organization and learn methods of implementing change, such as negotiation techniques and motivation. The course includes in-depth discussions of leadership skills, communication, conflict resolution, and goal integration. Students research a biotechnology organization and analyze what is working and not working within the management systems and suggest alternatives.

Marketing Aspects of Biotechnology

This course introduces students to the strategic and tactical approaches used in the marketing of biotechnological products and services. Students gain a thorough understanding of the research and planning necessary to develop a marketing plan, the relationship between the marketing and sales functions, the difference between marketing a scientific product and a scientific service, pricing strategies, distribution alternatives, communications, promotion, and the importance of perception. Knowledge of marketing terminology and techniques prove helpful to anyone in the industry.

Creating a Biotechnology Enterprise

This course provides a foundation to start or help grow a young biotechnology company from inception through early growth. Topics addressed shall include market assessment of innovative technology, patents and licensing, corporate law, preparing a business plan, raising money from angels and venture capitalists, government grants, strategic alliances, sales and marketing, real estate, human resources, and regulatory affairs. The course provides a survey and overview of the key tasks and challenges typically faced by biotech entrepreneurs, their management team, and directors. Students will prepare a business plan for a biotech start-up and present the plan to a panel of industry experts and financiers. Leaders from our local bioscience community will be guest lecturers for many of the classes.

Research Ethics

This course covers fundamental concepts in ethical decision-making in research through discussions of case studies, videos, and film. Students explore issues central to these areas, such as the appropriate use of animals in research, protections against risk in human subjects research, and topics related to integrity of the research process such as: authorship, peer review, and the ethics of the business of science.

Regulatory Processes for Domestic and Global Biotechnology Products

Regulatory affairs (RA) comprises the rules and regulations governing product development and post-approval marketing. In the U.S., the FDA establishes and oversees the applicable regulations. Biotechnology products may be classified as drugs, biologics, medical devices, or combination products. Each type is regulated by a different center within the FDA. This course provides an overview of RA and its impact on product development. Topics include RA history, regulatory agencies, how to access regulatory information, requirements for submissions and approval for drugs, biologics, and medical devices; compliance with GXP (Good Laboratory/Clinical/Manufacturing Practices), Quality Systems Regulations, ethics, and FDA inspections.

Legal Aspects of Biotechnology

In this course students gain a thorough understanding of the legal steps necessary to protect and market biotechnological inventions and of the procedures required to obtain the necessary permits and licenses from government agencies. Topics include inventorship and ownership issues in academia and industry; what can and should be patented in the United States and in other countries; how patents are granted; how to avoid losing patent rights; how to enforce and defend patents; and how to transfer rights to technology.

Bioscience Communication

Researchers must communicate effectively so their discoveries can be shared with others. In this course, students learn how to communicate their ideas to other researchers, their scientific peers, and investment communities. Students master both written and verbal communication skills; they hone their expertise at making both formal and informal oral presentations; they prepare poster presentations, and develop their own public speaking strategies. The course also presents personal strategies for improving daily communications, cross-cultural communications, and non-verbal skills. Students improve their written communication, editing, and informal writing skills. Participants also learn effective e-mail strategies for getting their message across, and learn how effective writing can improve their chances of getting grant applications approved. Class assignments include preparation of scientific papers, general science writing, oral presentations, PowerPoint presentations, and scientific posters.

Biological Processes in Regulatory Affairs

This course provides an overview of the biological processes and laboratory techniques utilized for the discovery, development and evaluation of therapeutic drugs. Students investigate drug development processes such as gene cloning, culture scale-up, downstream processing, and product purification. Emphasis is placed on the theory and application of laboratory methods used in drug development, such as recombinant DNA techniques, antibody technology, protein purification, immunoassays, high-throughput drug screening, chromatography, electrophoresis, cell receptor characterization, pharmacokinetics, drug toxicity testing and evaluation of therapeutic drugs, diagnostics, and vaccines.

International Regulatory Law

The sale of food, drugs, or medical devices generally requires prior government approval of the product and how the product will be made and sold—for example, what representations will be made on the package and the label. The legal parameters governing the approval process vary across the world, and the regulatory rules and climate in, for example, Europe, Asia, Canada, and Australia will be examined. Those procedures then will be compared and contrasted with the laws, regulations, and processes of the United States. Opportunities for collaboration and cooperation between and among countries, such as settling on guidelines for conducting clinical trials in a foreign country, will be discussed. In light of the globalization of economies, is there a possibility of harmonization of approval processes? The regulation of the importation and exportation of products and information across borders will conclude the course.

International Regulatory Affairs

The pharmaceutical/biotechnology product approval and marketing requires a good understanding of international regulatory affairs in order to successfully compete in today's global marketplace. It is important for tomorrow's leaders to understand and follow the regulatory differences to ensure optimum product development strategies, regulatory approvals and designs for exports conforming to the foreign regulatory bodies. There are various product development strategies that industry is using to shorten the product development time by conducting preclinical programs outside the United States , however, the strategy requires a careful planning and interaction with the U.S. and foreign regulatory agencies. With the increase in globalization of economy and exports, international regulations will have a bigger impact on the biotechnology business in the future. The course provides a review and analysis of the pharmaceutical/ biotechnology product approval processes within the world's major markets. The key strategies required in preclinical product development to marketing approval of the products in Europe , Japan , and the United States will be compared and discussed. Students will explore the European Union (EU) regulations and their overall importance on international markets. The course will cover the salient features of common technical and regulatory documents required for submission and approval to the leading regulatory bodies in the world, general guidance documents, International Harmonization, and General Agreement on Tariffs and Trade (GATT).

Food and Drug Law

The Food, Drug, and Cosmetic Act (FDCA) governs the regulatory approval process for bringing a drug, food, or cosmetic to market. The class will begin with discussion of the administrative procedures followed by the FDA. The course will focus on the drug and medical device approval process, but will include a brief discussion on the food approval process, what constitutes generally regarded as safe (GRAS) and the food additive petition. Application for a new drug will be described and contrasted with the abbreviated new drug application process. Over-the-counter drugs are regulated in another manner, and the differences will be highlighted. Advertising and labeling of the drug also must comport with FDA guidelines. Good laboratory and manufacturing practices will be discussed, as will the inspections to ensure compliance. Students then will be exposed to the enforcement activities of the FDA, including searches, seizure actions, injunctions, criminal prosecutions, and civil penalties authorized under the FDC Act, as well as other statutes. The course will close with a discussion on the international movement of drugs, particularly in light of the varying costs of ethical pharmaceuticals.

Preparing a Successful Submission

In the United States , the FDA clears products to market through submissions by sponsors. These submissions are the pathway to market for regulated drugs, devices, biologics, and well-defined biotechnology products. Each submission type is reviewed by one or more of several different centers within the FDA. This course provides an overview of these submissions and the effect of submission type on product development. Topics include biological and biotech submissions as well as drug and medical device submissions, the review procedures and approval standards applied to each type, the guidance that applies to these submissions and any FDA inspections required for approval.

Marketing in a Regulated Environment

This course is designed to help students understand, use, and comply with the laws, regulations, and policy guidance documents governing advertising and promotion of products regulated by the Food and Drug Administration (FDA). The course provides an overview of the regulations, discussion of the FDA's advertising oversight, and FDA enforcement activities. The course content introduces students to advertising approved products. It provided insights on comparing competitor products, the need for head-to-head comparative data, dietary supplements, Rx products, OTC products, unapproved investigational products, and unapproved research products. The course also covers the different regulations from FDA, FTC, and CPSC. Labeling requirements, DoT, OSHA, and other international advertising activities are also covered.

Practicum in Bioscience Regulatory Affairs (MS BRA Only)

This integrative case-based course will focus on applying knowledge gained from previous courses in the MS Bioscience Regulatory Affairs program to actual cases from the U.S. Food and Drug Administration. For each case, students will assume the role of either a regulatory specialist, an FDA reviewer or senior-level policy-maker, or other involved stakeholders, such as a consumer group or an advocacy group. Students will be expected to research, evaluate, and present scientifically and legally justifiable positions on case studies from the perspective of their assigned roles. Students will present their perspectives to the class and be asked to debate the issues with the other students from the perspective of their assigned roles. The major responsibility of the students in this course will be to make scientifically and legally defensible recommendations and to justify them through oral and written communication.

Managerial Finance for Biotechnology I (formerly Financial Development for Biotechnology)

This course integrates the tools of financial analysis with real-world problems in the technology industries. Topics include modeling, costs and benefits, and ratio and break-even analysis. Students will read, prepare, and analyze financial statements. The difference between management financial and tax financial statements will also be covered.

Commercializing Biotechnology

This advanced course provides an integrated and practical

approach to considering the principal areas of concern an entity faces when commercializing biotechnology, from creating or obtaining the technology through partnering with others to further develop and commercialize the technology, and finally selling the business or business line that incorporates that technology. The focus of this course is to highlight key junctures in a biotechnology company's evolution; help students identify key financial, management, and business issues at those junctures; and present practical alternatives for students to consider to resolve those issues. This course builds upon 410.650 Legal Aspects of Biotechnology and 410.646 Creating a Biotechnology Enterprise, but they are not prerequisites for the course.

Validation in Biotechnology

Understanding validation and applying a comprehensive validation philosophy are essential in today's biotechnology industry. First and foremost, validation allows a company to operate in compliance with the regulations and guidance set forth by FDA. Perhaps more importantly, it results in equipment, assays, and processes that are well understood and robust, less prone to failure, and more cost-effective. This course will introduce the fundamentals of validation, validation master planning, resource management, types of validation and the associated documentation, departmental roles and interaction, and the differences between commissioning and validation. Students will have an opportunity to solve real-world problems, generate actual validation documents, and develop validation program elements that balance regulatory requirements, operational needs, and business expectations.

Introduction to cGMP Compliance

There are many sources describing Current Good Manufacturing Practice. The fundamentals are published in the Federal Food, Drug and Cosmetic Act and in 21 CFR part 211. These regulations, however, only begin to describe the practices required in the pharmaceutical and biotech industries. Additional sources include guidance published by the Food and Drug Administration, EIRs, Form 483s, warning letters and podium comments from FDA speakers. Students will learn the history of the regulations and the Food and Drug Administration. Students will also learn practical solutions to the regulatory issues faced in the pharmaceutical and biotech industries today.

Technology Transfer and Commercialization

This course is an introduction to the multidisciplinary aspects involved in the process of bringing technical developments, particularly research emanating from universities and other nonprofit organizations, into commercial use. The course will provide an overview of the key policies, concepts, tools, issues, practices, and trends that are shaping the technology transfer field, with an emphasis on the life sciences sector.

Emerging Issues in Biotechnology

Biotechnology impacts the world and our social, political, and physical environment in ways many both inside and outside the industry may not fully understand or appreciate. It is critical to ensure that advances in biotechnology be accompanied by important public, political, and social considerations and discussions. This course will cover issues including domestic and global public perception of biotechnology, its benefits and risks, advances in bioagriculture and genetically modified foods, the impact of recombinant therapeutics on the pharmaceutical and healthcare industry, ways in which advances in biotechnology have and will continue to change our views of what life is, and how the political climate impacts advances in biotechnology discoveries. This highly interactive course will include thought-provoking debate and discussion with industry leaders, both proponents and opponents of biotechnology.

QA/QC for the Pharmaceutical and Biotechnology Industries

Many new quality initiatives for drugs, biotech products, and medical devices recently have been introduced. These include risk-based, science-based and systems-based assessments. Students will be presented with a comprehensive overview of the current best practices in quality assurance and quality control. Students will also be exposed to the most recent theories and expectations from the Food and Drug Administration.

Ethical, Legal, and Regulatory Aspects of the Biotechnology Enterprise

This course provides an overview of the important ethical, legal, and regulatory issues that are critical to the biotechnology industry. The course is divided into the three sections of ethics, legal issues, and regulations, each section taught by a faculty member who specializes in the area. The three sections are organized to provide basic material in a way that allows for an appreciation of how each influences the others. The section on ethics introduces students to core ethical values that guide the practice of science in the biotechnology industry. The section on legal issues focuses on key legal concepts, such as protecting inventions and intellectual property, and licensing. The section on regulation addresses the range of regulatory oversight mechanisms with which the biotech industry must comply and strategies for doing so.

Leading Change in Bioscience

As bioscience companies grow and mature, leadership

needs evolve. Students will learn how to identify their company's position in the "Leadership Life Cycle" and learn how to select the right leadership capabilities based on their current organizational needs. Students will perform research to identify how the right leaders at the right time can dramatically improve organizational success. Class discussions will address the leadership needs of organizations from early stage research-based companies through fully integrated bio-pharmaceuticals. Through this course, students will be able to determine the stage a company or organization is in and will be able to recommend the critical qualities an ideal leadership team will have to ensure current and future growth and organizational success.

Drugs, Medical Devices, and Government

In bringing a food, drug, or medical device to market, Patent Office activities and Food and Drug Administration activities will come into play. The course will explore the administrative roles of each agency and those instances where an intersection or overlap of activities of the agencies occurs. For example, the patentability requirements of utility and enablement of therapeutic methods can depend on the onset of clinical trials. A major nexus occurs in the regulatory approval of a generic form of a patented ethical or "name brand" drug. The patent holder can have the term of the patent extended as compensation for the often prolonged time to obtain regulatory approval. Thus, the patent term is extended beyond the 17-year or 20-year term. Conversely, the generic manufacturer can enter the regulatory process before the patent term expires. Thus, the generic manufacturer legally can make and use the patented invention of another. That is an excused infringement. The earliest generic manufacturer that files an ANDA also may be granted a monopoly period for exclusive sale. The practical aspects associated with the interplay of those two activities will be explored as will the ethical and public policy issues raised by the two activities.

Crisis Management

This course will provide students with a fundamental understanding of crisis management, risk communications, media relations and public opinion research techniques in the context of the worldwide controversy over biotechnology. Students will be introduced to crisis management principles, strategies, tactics, and communications methods that will enable them to predict, manage, and control real-world controversies they may confront as they pursue careers in biotechnology. Course participants will work as a team to develop a biotechnology-specific crisis management plan for analysis and discussion, and will also have the opportunity to hone their communications skills by participating in practice media interviews during in-class sessions.

Non-Science Elective Courses

Proseminar Biotechnology

The Biotechnology Proseminar introduces students to issues and challenges facing leaders of public and private sector organizations, and communities seeking to achieve shared goals within the biotechnology industry. The course brings together diverse academic science and business disciplines (science, regulatory affairs, marketing, finance, legal, ethics, communications, etc). It explores how these disciplines can be used as powerful tools to create effective leadership, and productive collaborations within the industry, while improving managerial decision making. The Proseminar frames and integrates the combined MS.MBA Biotechnology content, methods, and tools of inquiry and analysis.

Biotechnology Education and Applications

Modern biotechnology techniques are currently being

used in many areas such as food, agriculture, health care, forestry, the environment, minerals, oil and gas recovery and industrial processes to develop new products and processes, and to modify existing ones. Summarized below are some of the ways that modern biotechnology techniques have been applied in these areas.

Food and Agriculture: One of the most extensive applications of biotechnology has been in agriculture. Biotechnology techniques such as recombinant DNA techniques and mutagenesis have been used to develop plants with novel traits. These traits include herbicide tolerance and pest, insect and virus resistance. Biotechnology techniques have been used to produce biopesticides which are toxic to targeted plant pests. Also, experiments into genetic modification of aquatic organisms such as salmon, for such novel traits as enhanced growth, have been carried out using recombinant DNA techniques.

Health Care: To date, applications of biotechnology in health care have focused on fighting disease using the human body's own 'weapons'. Biotechnology medicines and therapies synthesize proteins, enzymes, antibodies and other substances which occur naturally in the human body, to fight infections and diseases. However, biotechnology also uses other living organisms, that is, plant and animal cells, viruses and yeasts to help produce human medicines. There are four main areas in health care in which biotechnology is currently being used: medicines, vaccines, diagnostics and gene therapy. Several biotechnology medicines have been developed and approved to treat such diseases as anaemia, growth deficiency in children, hemophilia, diabetes and others. Biotechnology companies have also developed therapeutic products for infectious agents such as HIV, hepatitis B and influenza. Biotechnology is responsible for new vaccines such as the hepatitis B vaccine which was approved in the U.S. to fight the hepatitis B virus. Biotechnology diagnostics have been used to detect a wide variety of diseases and genetic conditions. For example, gene therapy is used diagnostically to screen donated blood to protect the blood supply from HIV and hepatitis. Biotechnology diagnostics are also used in home pregnancy tests. In a home pregnancy test kit, a protein called a monoclonal antibody (MAb), binds to Human Chorionic Gonadotrophin (HCG), causing a colour change. HCG is present in a woman's urine only during pregnancy. In gene therapy, genes are used as 'drugs' to treat hereditary genetic disorders. Faulty or missing genes can be replaced to prevent the occurrence of a genetic disease. Current gene therapy is primarily experiment-based, with a few human clinical trials, such as for the treatment of cystic fibrosis, in the early stages.

Environment: Biotechnology applications in the environment focus on using living organisms primarily to treat waste and prevent pollution. Examples of these applications include bio-filtration and bio-remediation. Bio-filtration refers to the use of microorganisms to remove complex pollutants from the air emissions and waste water discharges of many manufacturing processes. Bio-remediation refers to a number of processes that use living organisms to degrade toxic waste into harmless by-products such as water, carbon dioxide and other materials. Examples of bio-remediation processes include:

- *Bio-stimulation:* A technique which involves introducing nutrients to stimulate the growth of waste-eating microorganisms already present in the environment at a waste site.
- *Bio-augmentation:* A technique whereby microorganisms, not normally present in the

ecosystem of a contaminated site, are added to the site to clean up pollutants. The microorganisms added may be natural or genetically modified. The organisms die when their food is used up.

- *Phytoremediation:* A technique which uses certain plants and fungi, planted at waste sites, to naturally reclaim the sites without using chemical treatment, incineration or land filling. These plants and fungi are used because they flourish by accumulating metals and other waste materials present in the soil.

Forestry: To date, only a few biotechnology-derived products and processes used in forestry have been commercialized. Most biotechnology applications in forestry are still in the research and development stage. However, some commercial applications include the development of bacterial bio-pesticides such as *Bacillus thuringiensis* (Bt), as an alternative to chemical pesticides. Tissue culture technology is also being used to produce genetically modified seedlings for forest regeneration. For example, BCRI Inc., a biotechnology company in British Columbia, is using tissue culture to produce weevil-resistant Sitka spruce. Companies are also developing bio-fertilizers such as plant growth-promoting rhizobia bacterial fertilizers. In addition, genetic engineering has been used to develop traits of agronomic interest such as pest and disease resistance, within trees. The pulp and paper industry has used enzymes and microorganisms in their bio-bleaching and de-inking processes.

Industrial Processes: Biotechnology has been applied in a variety of industrial processes in different ways, particularly in the use of biocatalysts in manufacturing processes. Biocatalysts are substances that initiate or modify the rate of a biological process and are generally consumed in the process. Some examples of industrial processes where biotechnology has been applied include chemicals, starch/ grain processing, cleaning and textile industrial processes. In the chemicals industry, biotechnology has been used to produce commodity and specialty chemicals. It is also used in the starch and grain processing industries through the use of enzymes to turn starch into glucose and fructose. Corn and other grains can be converted to sweeteners such as high-fructose corn syrup and maltose syrup, using enzymes. Biotechnology applications have been used to produce ethanol from grain. Furthermore, biotechnology is used in the textile industry for the finishing of fabrics and garments. In the pulp and paper industry, bio-pulping is used to manufacture some products.

Metals and Minerals Recovery: Biotechnology has been used in the mining industry in such processes as bio-leaching and metal bio-remediation and recovery. Bio-leaching is the use of bacteria to extract valuable metals. Metal bio-remediation and recovery refers to the use of biologically created enzymes in the degreasing process to recover metals of value.

Energy: To date, biotechnology applications in the energy industry include the production of cleaner coal and petroleum by removing sulfur and thus reducing the environmental contaminants released during combustion. Biotechnology has also been used widely to produce bio fuels such as bio-ethanol and bio-diesel. Bio fuels are alcohols, ethers, esters and other organic chemicals made from biomass such as herbaceous and woody plants, agricultural and forestry residues and industrial waste. These fuels can be used for electricity and as fuels for transportation.

Human Applications: When treating problems that arise from genetic disorder, one solution is gene therapy. A genetic disorder is a situation where some genes are missing or faulty. When this happens, genes may be expressed in unfavorable ways or not at all, and this generally leads to further complications. The idea of gene therapy is that a non-pathogenic virus or other delivery system can be used to insert a piece of DNA—a good copy of the gene—into cells of the living individual. The modified cells would divide as normal and each division would produce cells that express the desired trait.

The result would be that he/she would then have the ability to express the trait that was previously absent at least partially. This form of genetic engineering could help alleviate many problems, such as diabetes, cystic fibrosis, or other genetic diseases. The potential of genetic engineering to cure medical conditions opens the question of exactly what such a condition is. Some advocates see aging and death as medical conditions and engineering problems to be solved. They see human genetic engineering as a key tool in this *(see life extension)*. The difference between cure and enhancement from their perspective is merely one of degree.

Theoretically genetic engineering could be used to drastically change people's genomes which could enable people to regrow limbs, the spine, the brain. It could also be used to make people stronger, faster, smarter, or to increase the capacity of the lungs, among other things. If a gene exists in nature, it could be brought over to a human cell. Others feel that there is an important distinction between using genetic technologies to treat those who are suffering and to make those who are already healthy superior to others. There is widespread agreement that germline engineering should not currently be allowed for either therapeutic and enhancement applications, as evidenced by a recent report by the American Association for the

Advancement of Science. Though theory and speculation suggest that genetic engineering could be used to make people stronger, faster, smarter, or to increase lung capacity, the AAAS report finds that there is little evidence that this currently can be done safely or without unethical human experiments.

"Transgenic organism" is now the preferred term for genetically modified organisms with extra-genome (foreign genetic) information, as opposed to "genetically engineered" or "genetically modified" organisms (which may refer to changes made within the genome such as amplification or deletion of genes). The first Genetically Engineered drug was human insulin approved by the USA's FDA in 1982. Another early application of GE was to create human growth hormone as replacement for a drug that was previously extracted from human cadavers. In 1986 the FDA approved the first genetically engineered vaccine for humans, for hepatitis B. Since these early uses of the technology in medicine the use of the GE has expanded to supply many drugs and vaccines. One of the best known applications of genetic engineering is that of the creation of genetically modified organisms (GMOs). There are potentially momentous biotechnological applications of GM, for example oral vaccines produced naturally in fruit, at very low cost. A radical ambition of some groups is human enhancement via genetics, eventually by molecular engineering. DNA sequencing is a technique which is used to identify each base in DNA. Although the costs of DNA sequencing has dropped dramatically, the NIH estimates it costs at least $10 million to sequence 3 billion base pairs - the size of the whole human genome. Although there has been a tremendous revolution in the biological sciences in the past twenty years, there is still a great deal that remains to be discovered. The completion of the sequencing of the human genome, as well as the genomes of most agriculturally and scientifically important plants and animals, has increased the possibilities of genetic research immeasurably. Expedient and inexpensive access to comprehensive genetic data has become a reality, with billions of sequenced nucleotides already online and annotated. Now that the rapid sequencing of arbitrarily large genomes has become a simple, if not trivial affair, a much greater challenge will be elucidating function of the extraordinarily complex web of interacting proteins, dubbed the proteome, that constitutes and powers all living things. Genetic engineering has become the gold standard in protein research, and major research progress has been made using a wide variety of techniques, including:

- Loss of function, such as in a knockout experiment, in which an organism is engineered to lack the activity of one or more genes. This allows the experimenter to analyze the defects caused by this mutation, and can be considerably useful in unearthing the function of a gene. It is used especially frequently in developmental biology. A knockout experiment involves the creation and manipulation of a DNA construct in vitro, which, in a simple knockout, consists of a copy of the desired gene which has been slightly altered such as to cripple its function. The construct is then taken up by embryonic stem cells, where the engineered copy of the gene replaces the organism's own gene. These stem cells are injected into blastocysts, which are implanted into surrogate mothers. Another method, useful in organisms such as Drosophila (fruit fly), is to induce mutations in a large population and then screen the progeny for the desired mutation. A similar process can be used in both plants and prokaryotes.

- Gain of function experiments, the logical counterpart of knockouts. These are sometimes performed in conjunction with knockout experiments to more finely establish the function of the desired gene. The process is much the same as that in knockout engineering, except that the construct is designed to increase the function of the gene, usually by providing extra copies of the gene or attracting more frequent transcription.
- 'Tracking' experiments, which seek to gain information about the localization and interaction of the desired protein. One way to do this is to replace the wild-type gene with a 'fusion' gene, which is a juxtaposition of the wild-type gene with a reporting element such as Green Fluorescent Protein (GFP) that will allow easy visualization of the products of the genetic modification. While this is a useful technique, the manipulation can destroy the function of the gene, creating secondary effects and possibly calling into question the results of the experiment. More sophisticated techniques are now in development that can track protein products without mitigating their function, such as the addition of small sequences which will serve as binding motifs to monoclonal antibodies.

REFERENCES

Cohen, S. N., Chang,A.C., Boyer, H.W. & Helling,R. B.

Collins, F. & Galas,D. A new five-year plan for the US Human Genome Project. *Science* 262, 43–46 (1993).

Collins, F. S. *et al.*New goals for the US Human Genome Project: 1998–2003. *Science* 282, 682–689 (1998).

Drews, J.& Ryser, S. The role of innovation in drug development.

Druker, B. J. Imatinib alone and in combination for chronic myeloid leukemia. *Semin.Hematol.* 40, 50–8 (2003).

Fuller, B. P. *et al.* Policy forum: Ethics — privacy in genetics research. *Science* 285, 1359–1361 (1999).

Gabriel, S. B. *et al.* The structure of haplotype blocks in the human genome. *Science* 296, 2225–2229 (2002).

Gardner, M. J. *et al.* Genome sequence of the human malaria parasite *Plasmodium falciparum*. *Nature* 419, 498–511 (2002).

Gavin, A. C. *et al*. Functional organization of the yeast proteome by systematic analysis of protein complexes. *Nature* 415, 141–147 (2002).

Golub, T.R.Genomic approaches to the pathogenesis of hematologic malignancy.*Curr.Opin.Hematol.* 8, 252–261 (2001).

Guttmacher and Mark S.Guyer are at the National Human Genome Research Institute, National Institutes of Health, Bethesda, Maryland 20892, USA.

Guttmacher,A. E.& Collins, F. S. Genomic medicine — A primer.*N.Engl. J.Med.* 347, 1512–1520 (2002).

Hirschhorn, J.N., Lohmueller, K., Byrne, E. & Hirschhorn,K.A comprehensive review of genetic association studies. *Genet.*

Holt, R. A. *et al.* The genome sequence of the malaria mosquito *Anopheles gambiae*. *Science* 298, 129–149 (2002).

Horii, T., Nagao, Y., Tokunaga, T., and Imai, H. 20003. Serum-free culture of murine primordial germ cells and embryonic germ cells. Theriogen. 59, 1257-1264.

Horii, T., Nagao, Y., Tokunaga, T., and Imai, H. 20003. Serum-free culture of murine primordial germ cells and embryonic germ cells. Theriogen. 59, 1257-1264.

Jackson,D. A., Symons, R. H. & Berg, P. Biochemical method for inserting new genetic information into DNA of Simian Virus 40: circular SV40 DNA molecules containing lambda phage genes and the galactose operon of *Escherichia coli*. *Proc.Natl Acad. Sci.*

Kaczorowski, D.J., Patterson, E.S., Jastromb, W.E., Shamblott, M.J. 2002. Review Article. Glucose-responsive insulin-producing cells from stem cells. Diabetes Metab.

Kawano H, Do YS, Kawano Y, Starnes V, Barr M, Law RE, Hsueh WA. Angiotensin II has multiple profibrotic effects in human cardiac fibroblasts. Circulation 2000;101:1130-1137.

Kellerman S, Moore JA, Zierhut W, Zimmer HG, Campbell J, Gerdes

AM. Nuclear DNA content and nucleation patterns in rat cardiac myocytes from different models of cardiac hypertrophy. *J Mol Cell Cardiol* , 497-505 (1992).

Kennedy M, Firpo M, Choi K, Wall C, Robertson S, Kabrun N, Keller G. A common precursor for primitive erythropoiesis and definitive haematopoiesis.

Kerr, D.A., Lladó, J., Shamblott, M.J., Maragakis, N.J., Irani, D.N., Crawford, T.O, Krishnan, C., Dike, S., Gearhart, J.D., and Rothstein, J.D. 2003. Human embryonic germ cell derivatives facilitate motor recovery of rats with diffuse motor neuron injury. J. Neurosci. 23, 5131-5140.

Kerr, D.A., Lladó, J., Shamblott, M.J., Maragakis, N.J., Irani, D.N., Crawford, T.O, Krishnan, C., Dike, S., Gearhart, J.D., and Rothstein, J.D. 2003. Human embryonic germ cell derivatives facilitate motor recovery of rats with diffuse motor neuron injury. J. Neurosci. 23, 5131-5140.

Kimura, T., Suzuki, A., Fujita, Y., Yomogida, K., Lomeli, H., Asada, N., Ikeuchi, M., Nagy, A., Mak, T.W., and Nakano, T. 2003. Development and Disease. Conditional loss of PTEN leads to testicular teratoma and enhances embryonic germ cell production. Development 130, 1691-1700.

Klug MG, Soonpaa MH, Koh GY, Field LJ (1996) Genetically selected cardiomyocytes from differentiating embryonic stem cells form stable intracardiac grafts. J Clin Invest 98:216-224.

Kocher AA, Schuster MD, Szabolcs MJ, Takuma S, Burkhoff D, Wang J, Homma S, Edwards NM, Itescu S. Neovascularization of ischemic myocardium by human bone-marrow-derived angioblasts prevents cardiomyocyte apoptosis, reduces remodeling and improves cardiac function. Nature Med.

Labastie M-C, Cortes F, Romeo P-H, Dulac C, Peault B. Molecular identity of hematopoietic precursor cells emerging in the hunian embryo. Blood 1998;92:3624-3635.

Liechty KW, MacKenzie TC, Shaaban AF, et al. Human mesenchymal stem cells engraft and demonstrate site-specific differentiation after in utero transplantation in sheep. Nat Med 2000, 6:1282-6.

Lin Y, Weisdorf DJ, Solovey A, Hebbel RP. Origins of circulating endothelial cells and endothelial outgrowth from blood. J Clin Invest 2000;105:71-77.

Lynch, H. T. & de la Chapelle, A. Genomic medicine: hereditary colorectal cancer.*N.Engl. J.Med.* 348, 919–932 (2003).

Lysaght, M.J., and Hazlehurst, A.L., "Private Sector Development of Stem Cell Technology and Therapeutic Cloning," *Tissue Engineering* 9(3): 555-561 (2003).

MacLellan WR and Schneider MD. Genetic dissection of cardiac growth control pathways Annu. Rev. Physiol. 2000. 62:289-320.

Mahon NG, O'Roke C, Codd MB, et al. Hospital mortality of acute myocardial infarction in the thrombolytic era. Heart 1999;81:478-82.

Maier S, Tertilt C, Chambron N, Gerauer K, Huser N, Heidecke C-D, Pfeffer K (2001) Inhibition of natural killer cells results in acceptance of cardiac allografts in CD28-/- mice. Nature Med 7:557-562.

Makino S, Fukuda K, Miyoshi S, et al. Cardiomyocytes can be generated from marrow stromal cells in vitro. J Clin Invest 1999, 103:697-705.

McCarthy PM, Rose EA, Macmanus Q, et al. Clinical experience with the Novacor ventricular assist system. J Thorac Cardiovasc Surg. 1991; 102:578-587.

McEwan PE, Gray GA, Sherry L, Webb DJ, Kenyon CJ. Differential effects of angiotensin II on cardiac cell proliferation and intramyocardial perivascular fibrosis in vivo. Circulation 1998;98:2765-2773.

Mendel, G.Versuche über Pflanzen-Hybriden.*Verhandlungen des naturforschenden Vereines, Abhandlungen, Brünn* 4, 3–47 (1866).

Migeon, B.R., Chowdry, A.K., Dunston, J.A., and McIntosh, I. 2001. Identification of *TSIX*, encoding an RNA antisense to human *XIST'*, reveals differences from its murine counterpart: Implications for X inactivation. Am. J. Human Genet. 69, 951-960.

Milki, A.A. et al., "Incidence of monozygotic twinning with blastocyst transfer compared to cleavage-stage transfer" *Fertility and Sterility* 79(3): 503-506 (2003).

Miller, P. S. Is there a pink slip in my genes? *J.Health Care Law Policy* 3, 225–265 (2000).

National Institutes of Health Revitalization Act of 1993, Pub. L. No. 103-43, § 121(c), 107 Stat. 122 (1993), repealing 45 C.F.R. § 46.204(d).

Nelissen-Vrancken H, Debets J, Snoeckx L, Daemen M, Smits J. Time-related normalization of maximal coronary flow in isolated perfused hearts of rats with myocardial infarction. Circulation 1996;93:349-355.

Nidorf SM, Siu SC, Galambos G, Weyman AE, Picard MH. Benefit of late coronary reperfusion on ventricular morphology and function after myocardial infarction. J Am Coll Cardiol 1992;20:307-313.

Norwitz, E. R., et al., "Implantation and the survival of early pregnancy," *New England Journal of Medicine* 345(19): 1400-1408 (2001)

Ogawa M, Kizumoto M, Nishikawa S, Fujimoto T, Kodama H, Nishikawa SI. Blood 1999;93:1168-1177.

Onyango, P., Jiang, S., Uejima, H., Shamblott, M.J., Gearhart, J.D., Cui, H., and Feinberg, A.P. 2002. Monoallelic expression and methylation of imprinted genes in human and mouse embryonic germ cell lineages. Proc. Natl. Acad. Sci. USA 99, 10599-10604.

Onyango, P., Jiang, S., Uejima, H., Shamblott, M.J., Gearhart, J.D., Cui, H., and Feinberg, A.P. 2002. Monoallelic expression and methylation of imprinted genes in human and mouse embryonic germ cell lineages. Proc. Natl. Acad. Sci. USA 99, 10599-10604.

Orlic D, Kajstura J, Chimenti S, et al. Bone marrow cells regenerate infarcted myocardium. Nature 2001, 410:701-705.

Oz MC, Argenziano M, Catanese KA, et al. Bridge experience with long-term implantable left ventricular assist devices. Are they an alternative to transplantation? Circulation. 1997; 95:1844-1852.

Park, T.S., Hong, Y.H., Kwon, S.C., Lim, J.M., and Han, J.Y. 2003. Birth of germline chimeras by transfer of chicken embryonic germ (EG) cells into recipient embryos.

Pfeffer JM, Pfeffer MA, Fletcher PJ, Braunwald E. Progressive ventricular remodeling in rat with myocardial infarction. Am J Physiol 1991; 260:H1406-14.

Pfeffer MA, Braunwald E, Moye LA, et al. Effect of captopril on mortality and morbidity in patients with left ventricular dysfunction after myocardial infarction. Results of the survival and ventricular enlargement trial. The SAVE investigators. N Engl J Med 1992;327:669-677.

Pitt B, Segal R, Martinez FA, et al. Randomised trial of losartan versus captopril in patients over 65 with heart failure (Evaluation of Losartan in the Elderly Study, ELITE). Lancet 1997;349:747-752.

Pittenger MF, Mackay AM, Beck SC, et al. Multilineage potential of adult human mesenchymal stem cells. Science 1999, 284:143-147.

Pouzet B, Ghostine S, Vilquin J-T, Garcin I, Scorsin M, Hagege AA, Duboc D, Schwartz K, Menasche P (2001) Is skeletal myoblast transplantation clinically relevant in the era of angiotensin-converting enzyme inhibitors? Circulation 104:I223-228.

Rafii S, Shapiro F, Rimarachin J, Nachman R, Ferris B, Weksler B, Moore AS, Asch AS. Isolation and characterization of human bone marrow microvascular endothelial cells: hematopoietic progenitor cell adhesion.

Ravichandran LV and Puvanakrishnan R. In vivo labeling studies on the biosynthesis and degradation of collagen in experimental myocardial myocardial infarction. Biochem Intl 1991;24:405-414.

Reich, D. E. & Lander, E. S. On the allelic spectrum of human disease. *Trends Genet.* 17, 502–510 (2001).

Rothenberg, K. *et al.* Genetic information and the workplace: Legislative approaches and policy challenges. *Science* 275, 1755–1757 (1997).

Sachidanandam, R. *et al.* A map of human genome sequence variation containing 1.42 million single nucleotide polymorphisms. *Nature* 409, 928–933 (2001).

Sanger, F.& Coulson, A. R. A rapid method for determining sequences in DNA by primed synthesis with DNA polymerase.

Selkoe, D. J.Alzheimer's disease: genes, proteins, and therapy. *Physiol. Rev.* 81, 741–66 (2001).

Shamblott, M.J., et al., "Derivation of pluripotent stem cells from cultured human primordial germ cells" *Proceedings of the National Academy of Sciences.* 95(23): 13726-13731 (1998).

Sidow, A. Sequence first.Ask questions later. *Cell* 111, 13–16 (2002).

Smith, L. M. *et al.* Fluorescence detection in automated DNAsequence analysis. *Nature* 321, 674–679 (1986).

Smythe GM, Grounds MD (2000) Exposure to tissue culture conditions can adversely affect myoblast behaviour in vivo in whole muscle grafts: implications for myoblast transfer therapy. Cell Transplant 9: 379-393.

Smythe G.M., Hodgetts S.I., Grounds M.D. (2001) Problems and solutions in myoblast transfer therapy. J Cell Mol Med 5:33-47.

Stockwell, B.R. Chemical genetics: Ligand-based discovery of gene function. *Nature Rev. Genet.* 1, 116–125 (2000).

2

Biotechnology Experiments: Upcoming Areas

PRODUCTION AND STABILITY STUDIES OF THE BIOEMULSIFIER

Evaluation of both tenso-active and emulsifying activities indicated that a biosurfactant was produced by the newly isolated and promising strain *Candida glabrata* isolated from mangrove sediments. The extracellular water-soluble emulsifying agent was isolated and identified as a heteropolymer. The maximum of bioemulsifier production was observed when the strain was grown on soluble and insoluble substrates cotton seed oil plus glucose, reaching values of 10.0 g/l after 144 hrs at 200 rpm. The cell-free culture broth containing the examined agent lowered the surface tension of the medium to 31 mN/m. Stable and compact emulsions with emulsifying activity of 75% of cotton seed oil were detected. The emulsification capacity remained practically unaltered within a wide pH (2-12), temperature (4-80ºC) ranges and under NaCl concentrations up to 10%.

Surfactants and emulsifiers are indispensable components of daily life. They are widely used in the pharmaceutical, cosmetic, petroleum and food industries (Makkar and Cameotra, 1998; Lang, 2002). The surfactant industry now exceeds US$ 9 billion per year (Desai and Banat, 1997). Most of these compounds are of petroleum origin, which are not easily biodegradable and their manufacturing processes and by-products can be environmentally hazardous. Increased environmental awareness and strict legislation has made environmental compatibility of surfactants an important factor in their applications for various uses (Maier and Soberon-Chavez, 2000). Several different microbial products that exhibit surface-active properties have been identified in the past. These so called biosurfactants are produced by certain bacteria and by a number of yeasts and filamentous fungi. They include low-molecular-weight glycolipids, lipopeptides and high-molecular-weight lipid-containing polymers such as lipoproteins, lipopolyssacharide-protein complexes and polysaccharide-protein-fatty acid complexes (Ron and Rosenberg, 2001). Because biosurfactants are readly biodegradable and can be produced in large amounts by microorganisms and thus are not dependent on petroleum-derived products, they might well be able to replace, in some instances, the traditional synthetic surfactants (Banat et al. 2000).

The success of biosurfactant production depends on the development of cheaper processes and the use of low cost raw materials, which account for 10-30% of the overall cost. Most oils and fats produced in the world are used in the food industry, which generates great quantities of wastes, tallow, lard, marine oils and free fatty acids from the extraction of seed oils (Makkar and Cameotra, 1998; Makkar and Cameotra, 2002). The literature shows that a wide range of carbon sources, including agricultural renewable resources, like sugars and oils, are suitable carbon sources for production of ecologically safe biosurfactants with good properties. (Gallert and Winter, 2002). Optimal yields of bioemulsifier are usually obtained when carbohydrate and vegetable oil are used as substrate (Zhou and Kosaric, 1995).

Among yeasts, *Candida* species have been widely employed for insoluble substrates fermentation and have been reported to produce surface active agents (Sarubbo et al. 1999; Sarubbo et al. 2001). The objective of this work is to investigate the production of a biosurfactant by *Candida glabrata* isolated from mangrove sediments and to show the capacity of the emulsifying agent in stabilizing water-in-oil emulsions.

Candida glabrata UCP 1002 was isolated from mangrove sediment collected in the City of Rio Formoso, Pernambuco State, Brazil by Gomes et al. (2000). The isolation and identification of strain are according to O'Donnell (1979). The organism was kindly supplied from the Culture Collection of Nucleous of Resource in Environmental Sciences, Catholic University of Pernambuco, Recife, PE, Brazil. The culture was maintained at 5ºC on Yeast Mold Agar (YMA) slants containing (w/v): yeast extract (0.3%), malt extract (0.3%), tryptone (0.5%), D-glucose (1%) and agar (5%). Transfers were made to fresh agar slants each month to maintain viability.

n-Hexadecane was obtained from Sigma Chemical Co. (St. Louis, MO); food grade cotton seed oil was kindly supplied from Bunge Alimentos S.A. (SC, Brasil). Other chemicals used were analytical grade.

Cultures were grown on a mineral medium containing 0.1% NH_4NO_3,0.02% KH_2PO_4, 0.02% $MgSO_4.7H_2O$, 0.3% yeast extract and 7.5% cotton seed oil plus 5.0% glucose as substrates. The final pH of the medium was 5.7.

The *Candida glabrata* was grown in solid medium at

27°C for 48-72 hrs; then, a loopful of the cream coloured culture was transferred to Erlenmeyer flasks of 250 ml containing 50 ml of the liquid medium, Yeast Mold Broth (CYM) and incubated for one day at 27°C on an orbital shaker (200 rpm). The CYM culture contained 10^4 cells/ml and was used to initiate growth in mineral medium using a 10% vol/vol inoculum. The production of the emulsifier was carried out in Erlenmeyer's flasks of 500 ml containing 100 ml of the production medium and shaking at 200 rpm for 144 hrs at 27°C. The pH of the media was not adjusted during cultivation. Growth of the culture was monitored by counts on Neubauer Camera and plate YMA. The fermentation's was monitored by aseptically removing of samples until the end of the experiment. All analyses were performed in triplicate and did not vary more than 5%.

Emulsification activity was measured using the method described by Cooper and Goldenberg (1987) whereby 6 ml of n-hexadecane or cotton seed oil was added to 4 ml of the culture broth free of cells in a graduated screwcap test tube and vortexed at 3000 rpm for 2 min. The emulsion stability was determined after 24 hrs, and the emulsification index was calculated by dividing the measured height of the emulsion layer by the mixture's total height and multiplying by 100. The efficiency of the biosurfactant as an emulsifying agent was measured on the cell-free broth. Variations between volumes of the oil phase (n-hexadecane or cotton seed oil) and aqueous phase were prepared for water in oil emulsions.

Stability studies were done using the cell-free broth obtained centrifuging the cultures at 10000 x g for 15 min. 4 ml of the culture broth free of cells were heated at 80°C, and cooled to room temperature, after which the emulsification activity was measured. The emulsification capacity of culture broth free of cells was also determined after exposure at lower temperature (0-4°C). To study the pH stability of the cell-free broth, the pH of the cell-free broth was adjusted to different pH values (2 to 12) and the emulsification activity was measured. The culture liquid pH was adjusted with 1 M NaOH. The effect of NaCl concentrations (2 to 10%) on the emulsification capacity of the culture broth free of cells was also determined.

Surface tension and critical micellar concentration (CMC) were determined on cell-free broth obtained by centrifuging the cultures at 10000 x g for 15 min with a Tensiometer model Sigma 70 (KSV Instruments LTD - Finland) using the Du Nouy ring method at room temperature. The CMC was determined by measuring the surface tensions of dilutions of cell-free broth in distilled water up to a constant value of surface tension. Measurements of surface tension from distilled water and from the mineral medium were used as controls.

The 144-h culture was refrigerated for 24 hrs to solidify the remaining oil and to effect yeast settling. The culture was filtered through Whatman No. 1 filter paper and centrifuged at 10000 x g for 15 min. The cell-free broth was concentrated (500 ml) by freeze drying and extracted three times with chloroform (1:1, by vol.) in a separatory funnel at 28°C.

Protein concentration in the isolated bioemulsifier was determined by the Lowry method (Lowry et al. 1951) using Bovine serum albumin as a standard. Carbohydrates were determined by the phenol-sulphuric acid method, using D-glucose as a standard (Hanson and Phillips, 1981). The lipid composition of the crude bioemulsifier was determined according to Manocha et al. (1980).

The biomass concentration, pH and emulsification index of *Candida glabrata* cultivation in mineral medium containing 7.5% of cotton seed oil plus 5% of glucose.

Maximum biomass concentration was achieved after 72 hrs. After 48 hrs of growth, a diauxic behaviour was observed, probably due to the consumption of other substrate used in the fermentation. During the exponential growth phase, culture medium pH gradually decreased from 5.7 to 2.6, after which it remained around 3.0. The profile of emulsification activity production was observed in three independently run fermentations. Emulsification of cotton seed oil increased with increasing biomass formation, reaching its optimum nearly at about 24 hrs, and after 48 hrs of growth it showed with constant values around 75% until the end of cultivation. Conversely, the emulsification of n-hexadecane started after the microorganism entered the stationary growth phase, with maximum activity of 67% after 96 hrs of cultivation.

For the measurements of surface tension of distilled water after different dilutions of the cell-free broth after 144 hrs of cultivation, it was found that the emulsifier agent obtained from glucose plus cotton seed oil could lower the surface tension of water (air-water interface) from 68 mN/m to 31 mN/m (CMC), which shows that it is a good surfactant. This result is similar to others surfactants produced by yeast from carbohydrate and vegetal oil as substrates (Davila et al. 1992; Zhou and Kosaric, 1995; Garcia-Ochoa and Casas, 1999). The efficiency of the bioemulsifier containing cell-free broth. The results showed that the biosurfactant form *Candida glabrata* was efficient in emulsificating the cotton seed oil once no significant variation in the emulsification index was observed for this substrate, however, for n-hexadecane emulsification, the reduction of cell-free broth volume decreased the emulsification capacity. The effect of added NaCl concentrations on n-hexadecane and cotton seed oil emulsification capacity of the cell-free broth. A reduction of approximately 20% of emulsification activity was observed with the addition of up to 10% (w/v) sodium chloride for both substrates, showing a relative tolerance over these salt concentrations. Once the salinity decreases the viscosity, it is possible that the increase of NaCl concentration had influenced the quality of the emulsion, thus reducing the emulsification capacity. Reductions in emulsification activity were also reported for other surfactants of microorganisms as *Candida lipolytica* grown in n-hexadecane (Cirigliano and Carman, 1984) and a mixed culture cultivated in molasses (Ghurye et al. 1994).

The effect of thermal treatment on the emulsifier activity of *Candada glabrata* culture showed that no appreciable changes in emulsification capacity occurred, if the cell-free broth was heated, once only 10% of activity was lost at 80°C. The lost of emulsifier activity could be explained by the denaturation of proteinaceous compounds of the bioemulsifier during heating. There was no significant change at lower temperature (4°C). Liposan from *Candida lipolytica* found to be relatively stable between 30 and 90°C, but lost 60% of its activity after boiling for 1 hr (Cirigliano and Carman, 1984).

The pH of the cell-free broth was varied from 2 to 12 to test the effect of pH on emulsification capacity. No appreciable effect on activity was observed along the pH range, although it was observed an increase at pH 12, especially for cotton seed oil emulsification. Extremes of pH could possibly transform less surface-active species into more active emulsifiers by increased ionization. The activity of the biosurfactant produced by *Bacillus subtillis* was also pH stable (Makkar and Cameotra, 1998), while the effectiviness of liposan as emulsifier was limited to the acid to neutral pH (Cirigliano and Carman, 1984).

The examined agent was isolated from the culture filtrate of *Candida glabrata*. The precipitate collected in the aqueous phase recovered 100% of the emulsification activity of n-hexadecane that was present in the culture filtrate, while the emulsification activity of the cotton seed oil increased 25%. The average yield of precipitate in the aqueous phase was approximately 10.0 g/l.Bioemulsifier production by yeast *Candida utilis* varied from 0.26 to 0.93 g/l and depended on process conditions (Shepherd et al. 1995), while the extracellular emulsifying agent from *Curvularia lunata* yielded 2.6 g/l (Paraszkiewicz et al. 2002). The emulsification activity of both the isolated biosurfactant and the cell-free remained stable for a reasonable period (more than 4 weeks) under low temperature and upon sterilization.

Bioshyntesis of biosurfactants from a variety of bacteria and yeasts has been reported (Davila et al. 1992; Zhou and Kosaric, 1995; Daniel et al. 1999; Lang and Wullbrandt, 1999), most commonly involving rhamno-lipids, trehalose and sophorose-lipids. These usually contain various hydroxy fatty acids and carbohydrates and are characterized by unique surfactant properties (Ron and Rosenberg, 2001). Preliminary analysis of the bioemulsifier produced by the new strain *Candida glabrata* indicated that it was a heteropolymer, which consisted of 45% protein, 20% lipid and 10% carbohydrate. Liposan, an extracellular emulsifier synthesized by *Candida lipolytica* was composed of 93% carbohydrate and 7% protein (Cirigliano and Carman, 1984). Other polymeric emulsifiers containing proteins, carbohydrates and lipids were also produced by *Candida lipolytica* when grown in babassu vegetal oil (Sarubbo et al. 1999) or glucose (Sarubbo et al. 2001) as the sole carbon sources.

The results obtained in this work show that this new strain of *Candida glabrata* represents a valuable source of new compounds with surface-active properties, with potential of application in different industries. Work is in progress to improve the production process of these compounds by using industrial wastes as substrates.

CHEMICAL MODIFICATION PROTOCOLS OF ASPERGILLUS NIGER AND SORPTION OF LEAD ION

Biomasses of five fungi species (*Aspergillus niger*, *Penicillium austurianum, Saccharomyces cerevisiae, Mucor arcindloides and Trichoderma reesi* were evaluated for their uptake of lead ion from aqueous solution using batch systems. Both dead and live fungal biomasses were comparatively studied for their adsorption efficiencies. The effect of pH as one of the primary factors that influences sorption efficiency of metal ions in solution was also studied. Modification of the fungal biomass with the least sorption capacity was carried out using a four-step procedure. Three different modifying agents under the optimised experimental conditions were used. The percentage uptake of lead ion by fungi species ranged from: *Aspergillus niger*: 6.71-64.95% and 66.91-95.27%; *Penicillium austurianum*: 44.47-98.85% and 75.57–94.21%; *Saccharomyces cerevisiae*: 52.61-88.68% and 61.20-89.95%; *Mucor arcindloides*: 83.78-93.13% and 62.91-97.65% and *Trichoderma reesi*: 52.52-80.70% and 35.31-88.13% for dead and live biomass respectively. The influence of pH on metal uptake was evaluated at pH of 2, 4, 6 and 7. A regular pattern of sorption efficiency among the dead fungi species with respect to pH was observed and the % adsorption decreased in the order pH 7 > 2 > 6 > 4 with the exception of *Mucor* (pH 7 > 2 > 4 > 6). Modified biomass of *Aspergillus niger* with oxalic acid, malic acid and ethylenediamine tetraacetic acid (EDTA) recorded 92.84%, 48.11% and 39.83% uptake of Pb respectively which correspond to 69.65%, 41.23% and 29.25% increase when compared to 28.18% of the unmodified biomass. These quantitative adsorptions demonstrate the potential application of modified biomass for the removal of Pb ion from aqueous solution.

The removal and remediation of toxic metals from aqueous streams is imperative especially in countries where potable water is in short supply such as South Africa. The presence of toxic metals such as Pb, As, Hg and Cd in the environment is of great concern due to their health implications. They are non-essential metals with no physiological benefits to man and wildlife (Tyler, 1981; Borgmann, 1983). Over the years, anthropogenic activities have greatly increased the level of these toxic metals especially in aqueous systems, hence efficient and cost effective methods are being sought for their removal.

Conventional physical and chemical methods of metal removal from aqueous streams that have been applied include chemical precipitation, ion-exchange, adsorption and electrochemical techniques (Reed and Nonavinakere, 1992; Atwood et al. 2002; Di Palma et al. 2003). These methods are plagued with incomplete removal, involve the use of large volume of chemicals, high costs especially when contaminant concentrations are in the range of 10-100 mg l^{-1}, laborious and generate other wastes that require further disposal (Peters et al. 1985; Rostami and Joodaki, 2002).

Recent advances towards effective treatment methods for the removal of heavy metals in aqueous media involve the use of microorganisms (Viraraghavan and Yan, 2003; Cañizares-Villanueva et al. 2004; Ma et al. 2004). Attention has been focused lately on the application of fungal species for metal removal from aqueous media (Huang and Huang, 1996; Zouboulis et al. 2004; González-Chávez et al. 2004). Fungal cells, both living and dead such as *Aspergillus, Penicillium, Rhizopus, and Saccharomyces* have been applied to metal removal from aqueous streams using either batch or continuous modes (Rostami and Joodaki, 2002; Matis et al.

2003; Park et al. 2005). However, the use of dead cells has received greater attention compared to living ones. This is due to the argument that dead cells do not have toxicity limitations, no requirement of growth and nutrient media and easy desorption of adsorbed metal ions.

Generally however, the efficiency of fungal cells as sorbents varies greatly and depends on the physiological states, surface properties of cells, pH and other physicochemical parameters of the metal solution (Sag and Kutsal, 1996). The ability of microbial cells to bind metals from aqueous solutions has been demonstrated to be a function of solution pH (Wilhelmi and Duncan, 1995). In addition, metal ion adsorption on both non-specific and specific sorbents is pH dependent, as the pH affects the availability of metal ions in solution (speciation), as well as the metal binding sites onto cell surface (Zouboulis et al. 2004).

In an attempt to increase sorption efficiencies, fungal biomasses have been modified using varieties of modifying agents (Brady and Duncan, 1994; Matis et al. 2003). Fungal cell walls are made up of polysaccharides and proteins which contain many functional groups such as carboxyl, hydroxyl, phosphates and amino groups that can bind with metal ions (Veglio and Beolchini, 1997). Hence, the performance of an adsorbent depends on its surface properties as governed by the chemical structure, hydrophobic and hydrophilic characters of the cell wall. Biosorption mechanisms generally involve passive sequestration of metals by cell wall through processes of adsorption, ion-exchange, complexation, etc.

Several modification protocols of fungal biomasses using different modifying reagents have been reported (Brady and Duncan, 1994; Jianlong, 2002; Matis et al. 2003; Deng and Ting, 2005). These protocols, however involve complex and time consuming processes such as long chemical reaction steps, centrifugation, dialysis, specific reaction conditions (reaction under nitrogen) and multiple washing and re-suspension of biomass which might even contribute to poor performance of the modified biomass.

In this study, live and dead fungal biomasses of five fungi species – *Aspergillus niger*, *Penicillium austurianum*, *Saccharomyces cerevisiae*, *Mucor arcindloides* and *Trichoderma reesi* were evaluated for their sorption efficiencies of lead ion from aqueous solution using batch systems. A toxic metal of environmental significance (lead) was used as the test metal. The effect of pH as one of the primary factors that influences sorption efficiency of metal ions in solution was also studied since it is a key parameter in most biological processes and controls the growth and/or the adsorption capacity of substances. In order to eliminate the lengthy steps outlined in some of the aforementioned protocols, a new four-step approach to the modification procedure of the biomass of *Aspergillus niger* was investigated. The four-step approach involved: surface modification of biomass with suitable agents. This was followed by drying, washing and finally drying. Modification of the biomass (*Aspergillus niger*) with the least sorption capacity was carried out using three simple and cost effective modifying agents under the optimised experimental conditions.

All reagents used were of analytical grades. 1000 mg l^{-1} of lead standard stock solution (AAS grade) was prepared from reagent grade $Pb(NO_3)_2$ purchased from Heyns Laboratory; Fonteinbleau, South Africa. Working standard solutions were prepared by dilution of the stock. Deionised water from Millipore Instrument Corporation was used throughout the study. Portable hand-held combo (HI 98830) from Hanna Inc. was used for pH measurements. A Perkin Elmer AA 3030 Flame Atomic Absorption Spectrometer (FAAS) equipped with automatic background corrector was used for all trace metal determinations.

The experiments described in this study were carried out using five different species of fungi. These are *Aspergillus niger* (TUTC 120), *Penicillium sp* (TUTC 077), *Saccharomyces cerevisiae* (TUTC 112), *Mucor circinelloides f. circinelloides* (CBS 106.18) and *Trichoderma reesi* (TUTC 101). They were maintained on Malt Extract Agar (MEA).

Each of the five fungal species were grown on MEA plates for approximately 4 days or until the plates were fully grown. A 10% inoculum of spores or cellswas transferred to 1l Erlenmeyer flasks containing the cultivation medium. For the transfer of fungal cultures, about 20 ml of the spore buffer solution (pH 6.0) was added to a Petri dish containing fungal growth. Sterile Dragalsky needle was used to transfer the spores or yeast cells into the buffer solution. The buffer solution, consisting of 0.05 M solution of KH_2PO_4 l and 8.0 x $10^{-4\,M}$ of Tween 80 (Polyoxyethylene Sorbitan Monooleate was adjusted to pH 6.0 with 1N NaOH and the solution sterilised by autoclaving at 121°C for 15 min.

After inoculation, the organisms were grown at 25°C and 140 rpm for three days. Each flask contained 300 ml sterile cultivation medium (pH 6.0) that consisted of (g l^{-1}): citric acid, 0.15; $(NH_4)_2SO_4$, 2.50; $MgSO_4.7H_2O$, 0.50; KH_2PO_4, 6.00; $CaCl_2.2H_2O$, 0.04; yeast extract, 3.00; and glucose, 20.00. The medium was prepared by placing 300 ml of distilled water in 2 l Erlenmeyer flask and then citric acid, $(NH_4)_2SO_4$, $MgSO_4.7H_2O$, KH_2PO_4, $CaCl_2.2H_2O$, yeast extract, and glucose were added. Additional 400 ml of distilled water was poured into the flask and the medium was aseptically dispensed into 1l Erlenmeyer flasks and sterilized by autoclaving at 121°C for 15 min.

After the cultivation was completed, the fungal cells were harvested by filtration through Whatman number 1 filter paper and washed with 100 ml of distilled water. Half of the cells were immediately frozen at -70°C after which all the cultures were transferred to a freeze dryer for 48 hrs. The fungal cells were removed from the freeze dryer and stored in a freezer at -70°C. The remaining half of the harvested cells were autoclaved to obtain dead biomass after which the same procedure was followed, namely freeze drying and storing of the cells. Freeze-dried fungal biomasses were pulverised in clean acid washed mortar and pestle and then passed through 0.45 μm sieve. All optimisations were based on the fine powdered biomass obtained.

Working standard of lead ion was prepared from 1000 mg l^{-1} stock solution. 100 ml of 100 mg l^{-1} of lead standard was placed in glass stopper 250 ml Erlenmeyer flask and the solution pH was adjusted accordingly. pH range of 2, 4, 6 and 7 were evaluated for their influence on metal adsorption. 0.3 g of the powdered biomass was added to the flask and the cell suspension was adequately mixed by incubating and agitating at 150 rpm for 24 hrs at 25°C. All experiments were carried out in triplicate.

Fungal suspension was filtered using 0.45 μm membrane filter and the remaining non-adsorbed lead ion in cell-free filtrate was determined using FAAS. Good linearity of the calibration standards was obtained and triplicate analysis of each sample was carried out together with blank analysis. The amount of metal taken up by the biomass was calculated as the difference between the initial and final concentration of the metal in the aqueous solution.

Modification of the biomass of *Aspergillus niger* was carried out using three simple and cost effective reagents, oxalic acid, malic acid and ethylenediamine tetraacetic acid (EDTA). About 1 g of powdered biomass were separately placed in three clean 10 ml beakers and slightly moistened with deionised water. 7 ml of 0.05 M solution of EDTA, 0.1 M solution of oxalic and malic acids were separately added to each of the three beakers respectively. The acid/slurry was dried in the oven at 50°C overnight. The modified dried biomasses were then washed with 20 ml of distilled water to remove unreacted acid until the pH of the filtrate was near neutral. The modified biomass were again dried at 50ºC for 6 hrs, allowed to cool and then gently stirred with glass rod to return them to powdery form. 0.3 g of the modified biomasses was taken through metal adsorption and determination protocols at pH 7 as described earlier.

In this study, batch experimental systems were used in the optimization of metal sorption by fungal biomasses with respect to pH. Several literatures have reported the optimization of parameters such as adsorption time, metal concentrations, duration of agitation, effects of other ions in solution etc with respect to adsorptive capacity and efficiency of biosorbents (Thomas et al. 2000; Aldrich and Feng, 2004; Iqbal et al. 2005). In many of these studies, similar patterns of experimental findings were reported with respect to these parameters. For example, the adsorption kinetics usually follow an initial rapid increase in sorption pattern followed by a slow process; biosorption capacity of individual metal ion is reduced in the presence of other metal ions, increase in metal removal as biomass concentration increases (Natarajan et al. 1998; Yin et al. 1999; Viraraghavan and Guangyu, 2003; Park et al. 2005).

In this study however, emphasis was placed on the influence of pH on metal adsorption by biosorbent. pH has been reported as a key parameter in most biological processes and controls the growth and/or the adsorption capacity of substances (Crist and Martin, 1999; Wei and Tobin, 2004). In addition, solution pH influences the adsorption of metals differently.

Results of sorption efficiency by both live and dead biomass of each of the evaluated fungi species. The percentage uptake of lead ion by *Aspergillus niger* ranged from 6.71-

64.95% and 66.91-95.27% for dead and live biomass respectively across optimised pH of 2, 4, 6 and 7. Metal uptake by the living biomass was low initially at pH 2 before rising to a maximum value and then dropped slightly at pH 6 and suddenly drops at pH 7. There was no particular pattern of sorption efficiency with respect to pH with dead biomass. The highest sorption of 95.27% (living biomass) and lowest value of 6.71% (dead biomass) both at pH 4 were obtained for this fungal species. Sorption of lead ion from aqueous solution by live *Aspergillus niger* was generally better than those of the dead counterpart.

Percentage uptake of Pb ions by *Penicillium austurianum* (Figure 3) ranged from 44.47-98.85% and 75.57-94.21% for dead and live biomass respectively across the optimised pH. The highest (98.85%) and lowest sorption (44.47%) by dead biomass of this fungus was recorded at pH 7 and 4 respectively. Adsorption with respect to pH was in the order 7 > 2 > 6 > 4 and 4 > 2 > 6 > 7 for non-living and living biomass respectively. Appreciable sorption capacities of (>70%) of lead ion by both live and dead biomass of *Penicillium austurianum* was obtained with the exception of 44.47% (dead biomass) at pH 2.

From the uptake of lead ion by *Saccharomyces cerevisiae* varied between 52.61-88.68% and 61.20-89.95% for dead and live biomasses respectively. Also, the highest sorption of 88.68% at pH 4 and 89.95% at pH 7 were obtained for the two forms of the biomass in the same order. Comparable sorption capacities by the two forms (88.68% and 89.95%) of lead ion were demonstrated and the influence of pH on metal uptake was in the order of 7 > 6 = 2 > 4 and 4 > 2 > 6 > 7 for the dead and live biomass respectively of this fungus.

The best and consistent uptake of lead ion by the fungal

biomass across the optimised pH was observed with *Mucor arcindloides* (Figure 5). The percentage uptake ranged from 83.78-93.13% and 62.91-97.65% for dead and live biomass respectively. For the dead biomass, the highest uptake was observed at pH 7 while for live biomass, it was at pH 2. A downward trend of metal sorption up the evaluated pH range was observed with the live biomass while the opposite trend except at pH 7 was noted with the dead biomass of this fungus. The effect of pH on sorption capacity by this fungus is in the order of 7 > 2 > 4 > 6 and 2 > 4 > 6 > 7 for dead and live biomass respectively. pH 7 appeared to be optimal and suitable for metal uptake by the dead biomass. In the efficiency of metal sorption by *Trichoderma reesi* was in the range of 52.52-80.70% and 35.31-88.13% also for the dead and live biomass respectively. Again, the highest uptake of lead for dead biomass was observed at pH 7 while that of live was at pH 2. A regular pattern of sorption efficiency among the dead fungi with respect to pH was observed. The % adsorption decreased in the order pH 7 > 2 > 6 > 4 with the exception of *Mucor* (pH 7 > 2 > 4 > 6). The highest sorption occurred at pH 7.

pH changes generally have significant effects on metal solubilities and sorption/desorption reactions *i.e.* speciation in acidic and basic environment. At low pH (<2), lead ion tend to form insoluble salts hence would be unavailable in solution for sorption by the biomass. Also, at high pH (>9), it forms precipitates *e.g.* lead (II) hydroxide in the presence of sodium hydroxide. Hence, the results obtained from the present study revealed optimal sorption of lead ion at pH 7.

The sorption of metal cations usually occur at pH close to neutral while desorption is usually carried out by contact with acidic solution (Guibal, 2004). Optimal sorptions of metal ion at pH 7 by fungal biomass have also been reported

(Zouboulis et al. 2003; Deng and Ting, 2005). In addition, competition between protons and metal ions for the same sites is known to occur particularly at low pH values (Fourest and Roux, 1992). Hence in this study, further optimisation of the modified biomass of *Aspergillus niger* was carried out at pH 7 since better sorption efficiencies were obtained earlier with dead biomass. Live fungal biomass generally recorded better sorption efficiency than the dead counterpart. For practical purposes of metal removal from aqueous media however, dead fungal biomass has always been preferred to the live ones. This is due to earlier mentioned reasons of no toxicity problems, easy desorption of adsorbed metals, reusability of the biomass and non possession of obnoxious odour after usage among other reasons for dead fungal biomass.

With increasing interest in the application of dead microbial biomass for metal remediation from contaminated water, it became interesting to optimise the dead fungal biomass of *Aspergillus niger* since it gave the least adsorption efficiency. Results of the percentage uptake of Pb ion by unmodified and modified biomass of dead *Aspergillus niger* using oxalic acid, malic acid and EDTA as modifying agents. Oxalic acid modified biomass recorded 69.65% increase in sorption efficiency when compared to that of the unmodified biomass. This performance is comparable to the 92% removal of lead ion from aqueous solution by modified fungal biomass reported by Liu et al. (2005). Modified biomass using EDTA recorded the lowest sorption when compared with the other two modifying reagents. This could be as a result of the bulkiness of this molecule and also due to the orientation of the hydroxyl and the amine groups within the EDTA structure during the modification. Both oxalic and malic acids possess simpler structures which tend toward some form of linear molecular orientation that could facilitate easy bonding with binding sites (especially amino groups) on the biomass thus increasing the number of sites for metal bonding. Hence, the increase in modifying capacity of the three reagents was in the order: oxalic acid > malic acid > EDTA which was according to the structural complexity of the compounds. Generally, increase in uptake was obtained with the three modifying agents when compared with the unmodified biomass. The percentage uptake of lead ion by modified and unmodified biomass obtained in this study was compared to similar studies. 92% uptake of metal ion by modified fungal biomass reported by Liu et al. (2005) was comparable to the uptake of 92.84% obtained in this study. The value obtained in this study was better than the uptake of 76.3% of Pb ion at optimal pH of 5 and 100 mg l^{-1} of metal loading (Deng and Ting, 2005) and 61.3% of Cr^{6+} at 100 mg l^{-1} of metal loading at pH 2, 25 mg mycelia of *Lentinus sajor - caju* per 7.5 ml reaction medium at 25°C (Bayramoglu et al. 2005) reported in similar studies.

Modification of the cell wall and biomass of microorganisms have been explored extensively in attempts to increase the sorption efficiency of metal ions by the biomass using several modifying agents (Brady and Duncan, 1994; Jianlong, 2002; Deng and Ting, 2005). Modification processes of fungal biomasses reported in many literatures are laborious with many reaction steps such as centrifugation, dialysis, usage of cross-linking agents and multiple washing and re-suspension process. Chemical modification of yeast cell wall reported by Brady and Duncan, (1994) involved multiple centrifugation, dialysis, specific reaction condition (under nitrogen gas at pH 6.8) and washing steps. Glutaraldehyde was used as a cross-linking agent in an effort to improve the sorption capacity of *Penicillium chrysogenum* during biomass modification using polyethylene imine (PEI) (Deng and Ting, 2005).

Jianlong, (2002) reported a glutaraldehyde fixation to *Saccharomyces cerevisiae* involving up to 3 days of reaction process followed by filtration, washing and drying for 10 hrs. The additional use of cross-linking agents increases the reaction time and cost of the modification. Although, the cross-linking of *A. nodosum* by divinyl sulfone and formaldehyde under acidic conditions was found to reduce the swelling of the biomass so as to be amenable for column adsorption; the cross-linkage also resulted in reduced capacity from 215 mg Cd/g for raw biomass to 149 mg Cd/g for the formaldehyde-cross linked biomass (Adrian et al. 1999). The use of cationic surfactants, hexadecyl-trimethylammonium bromide and dodecylamine in the modification of the biomass of *Penicillium chrysogenum* has been reported (Matis et al. 2003). This cationic surfactant replaces cations such Na^+, K^+, Mg^{2+}, and Ca^{2+} present on the biomass surface rendering the surface positive or less negative. This facilitates the sorption of oxyanions and the treatment was found to improve the efficiency of sorption of As (V) by the dead biomass of *P. chrysogenum*. However, hexadecyl-trimethylammonium bromide is a toxic and hazardous chemical with severe eyes, skin and respiratory tract irritation. It also affects the gastrointestinal tract with symptoms such as nausea, vomiting and diarrhoea if accidentally ingested. At this juncture, it is pertinent to mention that great emphasis and consideration should be placed in the selection of reagents to be used for biomass modification. In this way, the release of toxic chemicals to the environment would be avoided.

In the present study however, a new approach to biomass modification was explored. This involves the use of common, readily available, cheap and most importantly, environmental friendly modifying agents which are oxalic acid, malic acid and EDTA. The modification procedure involves addition of the modifying agent to the powdered biomass which was then dried at low heat of 50°C in oven overnight (12 hrs). The biomass is washed using distilled water to remove excess reagent and finally dried at same temperature for a few hours (3-4 hrs). The protocol was found to be simple, fast and efficient. Efficiency and applicability of the modified biomass was revealed by the quantitative adsorption of lead ion from aqueous solution compared with the unmodified biomass. Cost of the adsorbent is an important parameter for consideration in the choice of materials to be used for metal removal from aqueous systems. A sorbent can be assumed or referred to as "low cost" if minimal processing of the material is required, is quite abundant in nature, or is a by-product or waste material from industrial processes. *Aspergillus niger* fits into one of these conditions because it could readily be obtained as industrial waste resulting from citric acid production which makes its utilisation both cost effective and environmental friendly. In addition, the proposed modification protocols could be said to be economically viable and worthy given the significant difference in sorption efficiency of 92% of Pb ion for the modified biomass compared with 28.12% of the unmodified using oxalic acid as the modifying agent; a common, cheap and readily available reagent.

The ability of fungi species to sorb lead ion from aqueous solution through batch experiments was demonstrated. Quantitative adsorption of lead ion by both live and dead biomass of fungal species was obtained with the exception of dead biomass of *Aspergillus niger* (6.71%) at pH 2. A regular pattern of sorption efficiency among the dead fungi species with respect to pH was observed and the % adsorption decreased in the order pH 7 > 2 > 6 > 4 with the exception of *Mucor* (pH 7 > 2 > 4 > 6). A new approach to biomass modification protocol was developed which was found to be simple, fast and efficient with respect to sorption of lead ions from aqueous system. Modification with oxalic acid gave the best sorption efficiency relative to malic acid and EDTA when compared with 28.18% of the unmodified biomass. The modified biomass of *Aspergillus niger* and the simplified protocol demonstrate their potential industrial applications for the removal of metals from contaminated water systems.

3

Biotechnology Research Related Equipments

Human cells are almost ridiculously tiny and efficient. Every one of them has an entire genetic instruction set—three billion As, Cs, Gs and Ts—packed into a nucleus one hundredth of a millimeter across. And a few cents worth of carbon, nitrogen and phosphorus will keep millions of cells happy in a dish, reproducing themselves and their DNA data.

Incyte Pharmaceuticals, Inc. (Palo Alto, Calif.) also stores DNA data, but in a way that is somewhat less compact and far more likely to impress a computer geek. Ground-zero for Incyte is a highly air-conditioned room filled with scores of black, towering supercomputers worth up to a million dollars each.

Nine of the ten largest pharmaceutical companies in the world each pay Incyte ~\$5 million per year to look at the data on those computers. The money has brought an expanding workforce—from 160 to 675 employees in the last two and a half years—and a frantic effort to keep generating more and more information. "We're maybe a year away from going bankrupt if we don't pursue new technologies at any one point," says Tod Klingler, Director of Research Bioinformatics at Incyte.

The manpower and computer power is needed to tame an ever-growing mountain of DNA sequence information. "We now have 24 organisms completely sequenced we have their entire genetic message written down," says Temple Smith of Boston University. "This information allows us to get information about one organism and use that to understand another."

BIOCHIPS: MICROCHIPS AND MEDICINE JOIN FORCES

Paranoid ranting is a staple on the internet, and the topic of biochips is a particular favorite. Biochips—any microprocessor chips that can be used in biology—mean many things to many people. The web's preachers focus on a simple idea: an implanted chip that identifies each person uniquely, and can be used to track their location.

This wonderful idea—capable of locating lost children, downed soldiers, and wandering Alzheimers patients—sounds alarming even to many mild-mannered civil libertarians. To Living Truth Ministries of Austin, TX, it is no less than "a tremendous bonanza for the coming Antichrist!" Second Advent Ministries of San Bernardino, Calif., warns us that the US social security number will soon be replaced (at the orders of the "New World Order United Nations-controlled world government", no less) by a chip-encoded "18-digit numbering system [which] will consist of THREE separate sets of SIX DIGITS each: 6-6-6! Coincidental? No. Not if you believe Bible prophecy."

Reality is somewhat less alarming. A simple ID chip is already walking around in tens of thousands of individuals, but all of them are pets. Companies such as AVID (Norco, Calif.), Electronic ID, Inc. (Cleburne, Tx.), and Electronic Identification Devices, Ltd. (Santa Barbara, Calif.) sell both the chips and the detectors. The chips are the size of an uncooked grain of rice, small enough to be injected under the skin using a syringe needle. They respond to a signal from the detector, held just a few feet away, by transmitting out an identification number. This number is then compared to database listings of registered pets.

Daniel Man, a plastic surgeon in private practice in Boca Raton, Fla., holds a patent on a more sinister device: a chip that would enable lost humans to be tracked by satellite. He has a large-scale model, but for miniaturization and regulatory approval he will need a substantial amount of cash. "There are some companies that are interested, but nothing is official," he says. "We're not giving up on it yet."

The civil liberties debate over biochips has obscured their more ethically benign and medically useful applications. Medical researchers have been working to integrate chips and people for many years, often plucking devices from well-known electronic appliances. Jeffrey Hausdorff of the Beth Israel Deaconess Medical Center (Boston, Mass.) has used the type of pressure-sensitive resistors found in the buttons of a microwave oven as stride timers. He places one sensor in the heel of a shoe, and one in the toe, and adds a computer to the ankle to calculate the duration of each stride.

"Young, healthy subjects can regulate the duration of each step very accurately," he says. But elderly patients prone to frequent falls have extremely variable stride times, a flag that could indicate the need for more strengthening exercises or a change in medication. Hausdorff is also using the system to determine the success of a treatment for congestive heart failure. By monitoring the numbers of strides that a person takes, he can directly measure the patient's activity level, bypassing the often flawed estimate made by the patient.

Hausdorff's chips are external, but a chip proposed by Sensors for Medicine and Science, Inc. (S4MS; Germantown, Md.) will be injected under the skin. The chip will allow diabetics to easily monitor the level of the sugar glucose in their blood. Diabetics currently use a skin prick and a hand-held blood test, and then medicate themselves with insulin depending on the result. The system is simple and works well, but the need to draw blood means that most diabetics don't test themselves as often as they should. Although they may get away with this in the short term, in later life those who monitored infrequently suffer from blindness, loss of circulation, and other complications.

The solution is more frequent testing, using a less invasive method. Various companies are trying to extract the required body fluid, using either an electric current (Cygnus, Inc., Redwood City, Calif.) or a tiny pore induced by a laser (SpectRx, Norcross, Ga.) or a small needle (Integ, St Paul, Minn.).

"All of these approaches ameliorate the drawing of blood," says George Daniloff, Chief Scientific Officer of S4MS, "but they still require manipulation of skin, and the skin may be perturbed." The S4MS chip, however, will sit underneath the skin, sense the glucose level, and send the result back out by radio-frequency communication.

A light-emitting diode (LED) starts off the detection process. The light that it produces hits a fluorescent chemical: one that absorbs incoming light and re-emits it at a longer wavelength. The longer wavelength of light is then detected, and the result is sent to a control panel outside the body.

Glucose is detected because the sugar reduces the amount of light that the fluorescent chemical re-emits. The more glucose there is, the less light that is detected. S4MS is still developing the perfect fluorescent chemical, but the key design innovation of the S4MS chip has been fully worked out. The idea is simple: the LED is sitting in a sea of the fluorescent molecules. In most detectors the light source is far away from the fluorescent molecules, and the inefficiencies that come with that mean more power and larger devices. The prototype S4MS chip uses a 22μW LED, almost forty times less powerful than the tiny power-on buttons on a computer keyboard. The low power requirements mean that energy can be supplied from the outside, by a process called induction. The fluorescent detection itself does not consume any chemicals or proteins, so the device is self sustaining.

Daniloff has a working model of an oxygen sensor that uses the same layout. With it's current circuitry it is about the size of a large shirt-button, but the final silicon wafer will be less than a millimeter square. "The scale transition is not a challenge," says Daniloff. "The challenge is to illustrate the principle." The oxygen sensor will be useful not only to monitor breathing in intensive care units, but also to check that packages of food, or containers of semiconductors stored under nitrogen gas, remain airtight.

Tom Ferrell of the Oak Ridge National Laboratory in Oak Ridge, Tenn., is also developing an oxygen-sensing chip, but his version sends light pulses out into the body. The light is absorbed to varying extents, depending on how much oxygen is being carried in the blood, and Ferrell's chip detects the light that is left. The rushes of blood pumped by the heart are also detected, so the same chip is a pulse monitor. A number of companies already make large-scale versions of such detectors, and Ferrell says he has "collaborated with almost every company in this area," although he will not name names.

Ferrell estimates that the oxygen chip is two years away, but he has already reduced the dimensions of his temperature-sensing chip to 3mm per side. The transition of certain semiconductors to their conducting state is inherently sensitive to temperature, so designing the sensor was simple enough. With some miniature radio-frequency transmitters, and foam-rubber earplugs to hold the chip in place, the device is complete. Ferrell sees applications for anyone from sick children, to chemotherapy patients who can be plagued by sudden rises in body temperature in response to their anti-cancer drugs.

Brain Surgery with an On-off Switch

Sensing and measuring is one thing, but can we switch the body on and off? Heart pacemakers use the crude approach: large jolts of electricity to synchronize the pumping of the heart. The electric pulses of the Activa implant, made by Medtronic, Inc. (Minneapolis, Minn.), are directed not at the heart but the brain. They turn off brain signals that cause the uncontrolled movements, or tremors, associated with diseases such as Parkinson's.

Drug therapy for Parkinson's disease aims to replace the brain messenger dopamine, a product of the brain cells that are dying. But eventually the drug's effects wear off, and the erratic movements come charging back. "I've had patients tell me that they scare their grandkids because they are flaying around like a fish," says Gary Heit, a neurosurgeon at Stanford University Medical Center. "Imagine your whole body twisted like a charleyhorse."

The next option is to remove a portion of the thalamus, a brain structure that triggers the tremors. The implant, cleared for use in the United States in August 1997, is a new alternative that uses high-frequency electrical pulses to reversibly shut off the thalamus. "It has the same benefits as thalamotomy [the brain operation] without the risks," says Heit. The implantation surgery is far less traumatic than thalamotomy, and if there are any post-operative problems the stimulator can simply be turned off.

Adding Back the Senses

The Activa implant interferes with aberrant brain functioning. The most ambitious bioengineers are trying to add back brain functions, restoring sight and sound where there was darkness and silence.

The success story in this field is the cochlear implant. Most hearing aids are glorified amplifiers, but the cochlear implant is for patients who have lost the hair cells that detect sound waves. For these individuals, no amount of amplification is enough.

The cochlear implant delivers electrical pulses directly to the nerve cells in the cochlea, the spiral-shaped structure that translates sound into nerve pulses. In normal hearing individuals, sound waves set up vibrations in the walls of the cochlea, and hair cells detect these vibrations. High frequency noises (deep notes) vibrate the base of the cochlea, while low frequency notes vibrate nearer the top of the spiral. The implant mimics the job of the hair cells. It splits the frequencies of incoming noises into a number of channels (typically eight), and then stimulates the appropriate part of the cochlea.

The two most successful implants are the Clarion (developed at the University of California at San Francisco (UCSF), Research Triangle Institute, N.C., and Advanced Bionics Corporation of Sylmar, Calif.) and the Nucleus (developed at the University of Melbourne, Australia, and made by Cochlear of Sydney, Australia). Upgrades largely focus on improving the speech processing software, which is operated by a minicomputer worn on the patient's belt. Theoretically, increasing the number of channels (and electrodes) could improve sound perception. But speech is perceived in an area of the cochlea only 14 mm long, and spacing the electrodes too close to each other causes signals to bleed from one channel to another.

The result is a broad-brush version of hearing. While some recipients of the devices report speech-like sounds, many characterize their new world as being populated with legions of quacking ducks or banging garbage cans. But the success is undeniable. "Currently two thirds to three quarters of patients [with more recent models] can understand speech without lip-reading," says Steve Rebscher, a member of the UCSF team. "It's pretty amazing, and certainly better than a lot of people anticipated these devices would do."

With the ear at least partially conquered, the next logical target is the eye. Several groups are working on implantable chips that mimic the action of photoreceptors, the light-sensing cells at the back of the eye. Photoreceptors are lost in both retinitis pigmentosa, a genetic disease, and age-related macular degeneration, the most common cause of lost sight in the developed world.

Joseph Rizzo of the Massachusetts Eye and Ear Infirmary, and John Wyatt of the Massachusetts Institute of Technology have made a twenty-electrode, 1mm-square chip, and implanted it at the back of rabbits' eyes. The original chip, the thickness of a human hair, put too much stress on the eye, so the new version is ten times thinner. The final set-up will include a fancy camera mounted on a pair of glasses. The camera will detect and encode the scene, then send it into the eye as a laser pulse, with the laser also providing the energy to drive the chip. Rizzo has confirmed that his tiny array of light receivers (photodiodes) can generate enough electricity to run the chip. He has also found that the amount of electricity needed to fire a nerve cell into action is ~100-fold lower in the eye than in the ear, so the currents can be smaller, and the electrodes more closely spaced.

For now the power supply comes from a wire inserted directly into the eye and, using this device, Rizzo has detected signals reaching the brain. "We know we can get visual information to the brain," he says, "but we don't know what kind of vision that will result in."

Eugene de Juan of Johns Hopkins Wilmer Eye Institute (Baltimore, Md.) is trying to answer that question by using human subjects. His electrodes, inserted directly into the eye, are large and somewhat crude. But his results have been startling. Completely blind patients have seen well-defined flashes, which change in position and brightness as de Juan changes the position of the electrode or the amount of current. In his most recent experiments, patients have identified simple shapes outlined by multiple electrodes. With as little as an 8 x8 array, de Juan believes he could approximate character recognition, and a 25 x 25 array might give a crude image.

The big money in eye implants is in Germany, where the government has pledged ~$10 million to two projects. One is similar to the US projects, in which chips are implanted on the surface of the retina, the structure at the back of the eye. The other project is putting its implants at the back of the retina, where the photoreceptors are normally found. These 'subretinal' chips may block the transport of oxygen and food to the overlying nerve cells, so Eberhart Zrenner of the University of Tübingen, Germany, is developing 'chain mail' electrode arrays, with plenty of holes for the delivery of supplies.

The subretinal approach more closely mimics the action of the lost photoreceptors. The chip delivers its message at the source, before intervening layers of retinal nerve cells do some complicated processing. Rizzo thinks that his chip's hardware will be able to compensate for these computations, but for now he has other things to worry about. "There are all sorts of processes that go on that we're just going to ignore initially," he says. As with the cochlear implant, he says, "the biological tests will show you the best way to stimulate. Ultimately the only thing that matters is what the person sees."

The first big test—showing that a remotely powered device can give true visual signals—is yet to be done. "We haven't got to first base yet," says Rizzo, "but we're close." With a working device many years away, Rizzo is a model of caution. "It's always better to talk down the possibilities than to talk them up, because there are many reasons why the implants might fail," he says. "This is science fiction stuff."

OSCP Biotechnology Team in USA

The Biotechnology Team located in EPA's Office of Science Coordination and Policy (OSCP) coordinates scientific, technical, and policy development activities within the Office of Prevention, Pesticides, and Toxic Substances (OPPTS). The OSCP Biotechnology Team is also a focal point for coordination with other Federal agencies on any issues involving biotechnology, including international activities.

BIOTECHNOLOGY CAPACITY BUILDING: A CASE STUDY

This report represents a summary of discussions that took place during a meeting held at FAO headquarters between 9th and 11th February 2005. The meeting addressed issues in connection with strengthening national plant breeding and biotechnology capacity. Over twenty representatives of agencies and institutions involved in genetic resource use and plant breeding, including an observer from the International Union for the Protection of New Varieties of Plants (UPOV) and about ten representatives from FAO attended the meeting. The participants came from approximately twenty countries.

Agriculture is the mainstay of the economies of numerous developing countries, supporting the livelihoods of large proportions of their populations. Subsistence agriculture is a particular feature of many of these countries, especially those that have suffered from the ravages of war and the vagaries of climate. With few alternatives, rural populations must often secure their sustenance and income from farming. Plant breeding, relying on judicious use of Plant Genetic Resources for Food and Agriculture (PGRFA), has supplied adapted crop varieties to many countries across the globe, with the principal exception of a number African countries. Green Revolution improved varieties have raised production levels, provided food security and generated income for many developing country farmers. The process of producing improved varieties is slow and requires long-term sustained commitment to plant breeding. In many countries application of biotechnology to agricultural research and production has hailed a new era. There is the promise that application of biotechnology can speed up some aspects of plant breeding, although the techniques complement rather than substitute for conventional plant breeding methods. However, neither the potential of biotechnology nor the proven value of plant breeding in general can be realised in the absence of adequate investment. Unfortunately public sector investment in plant breeding and biotechnology, and the research that supports them, has been in decline over recent times. National Agricultural Research Systems (NARS) have been the principal sufferers in this resource-constrained environment as plant breeding programmes have been trimmed and the possibilities for introducing biotechnological tools have receded. Private sector investment in agricultural research and plant breeding, often using the tools of biotechnology, has increased, but linkages with the public sector have been sub-optimal. Furthermore, scarce attention is paid to crops where seed trade is minimal, and which are often important crops of smallholder agriculture in developing countries. It has been suggested that realisation of the Millennium Development Goals might be facilitated by increased investment in plant breeding in both developed and developing nations. The declining capacity of many national plant breeding programmes has also meant that fewer crops can be researched and bred. There are many vitally important crops that have not received adequate attention from scientists and plant breeders. Given the current financial constraints, it seems unlikely they will receive the resources they merit to enable diversification for food security and poverty alleviation. The International Treaty (IT) on PGRFA, which has been ratified by 66 countries, serves to enhance access to and sustainable use of PGRFA. The Global Crop Diversity Fund was established to contribute to PGRFA conservation efforts. The issue remains, however, that conservation only represents half of the problem. While PGRFA conservation is crucial for providing the raw material for plant breeders to work with, PGRFA use is key to sustaining livelihoods, ensuring food security and contributing to national development.

The national plant breeding survey questionnaire

To quantify the status and progress in investment and production of national plant breeding programmes, FAO considered it necessary to implement a survey. The approach to the issue in general, including the questionnaire in particular, was designed through an expert consultation.

This in turn led to the paper Towards the Sustainable Use of Plant Genetic Resources Food and Agriculture (PGRFA): Strengthening Plant Breeding, which was prepared for the Working Group on PGRFA of the CGRFA (CGRFA/WG-PGR-2/03/2). The questionnaire is included. It is being used to gather information on plant breeding capacity in terms of budget allocation and staff establishment between 1985 and 2004. Moreover, it requires information to be supplied on the crops addressed in national programmes and the extent of application of biotechnology to plant breeding. At the time of the workshop, 18 countries had completed the questionnaire. It is planned to build on the results of the pilot phase and send the questionnaire to many more countries, with the ultimate aim of producing and maintaining an up-to-date database on national plant breeding programmes. Some of the information from the completed questionnaires was presented and discussed at the workshop.

Objectives and structure of the workshop

The objectives of this workshop were to bring together participants from developing countries of Africa, Asia and South America, representing a variety of backgrounds and expertise, to discuss the problem of the decline in plant breeding and how this might be remedied. The workshop aimed to provide guidance on, and detail linkages between, modern and traditional plant breeding. The necessary move from conservation of PGRFA to sustainable use was also a major discussion point.

The workshop programme is given a list of the workshop participants is provided. The workshop comprised presentations of country profiles, based on the results of the questionnaire. Discussion groups addressed many of the principal issues and constraints concerning plant breeding status and direction in developing countries. The presentations and discussions were synthesised into the following text and recommendations.

Preamble—A complex issue

Development of national plant breeding capacity and its current and future status is a complex issue. Geographical and economic circumstances differ enormously over the developed and developing world. Even on a regional basis, there can be large differences among countries. Civil conflict and natural disasters have had long-lasting effects on national capacities to undertake agricultural research. Countries also have very different natural resource bases; some are rich and others poor. Some have well-educated workforces while others do not. Given such complicating factors, it is not possible to describe a single model of the status of plant breeding and biotechnology application in the developing world. Nor is it possible to design a single set of remedial measures needed to halt the decline in plant breeding. Accurate models will necessarily be situation-specific.

Working group discussion

Presentations and discussions took place over the course of three days. For convenience, the main ideas and issues are discussed under six broad headings.

Public and private sectors: The public sector, in particular the NARS, has the largest responsibility for conducting agricultural research and applying the results. This includes producing improved crop varieties for cultivation by a country's farmers. The capacity to undertake this work varies enormously among countries and even within regions. This was highlighted during the presentations of country data and regional syntheses. While some countries have taken on board some of the more advanced methods of biotechnology, most are struggling to maintain, rehabilitate and initiate meaningful plant breeding programmes. Although the overall trend is towards reduced funding and decreased capacity to produce new, adapted varieties of essential crops, there are certain regions that are able to do a better job than others are. A few countries are very advanced. One of the principal factors underlying the problem of reduced funding faced by the public sector is lack of awareness among policy makers of the impact of plant breeding on national development. Those countries that were not able to benefit from the improved varieties and farming practices that derived from the Green Revolution have not made as much progress as those that were able to benefit. It is generally under-appreciated that the returns to investment in plant breeding are substantial and that without plant breeding it is difficult to effect progress in agrarian societies. The private sector encompasses a range of enterprises, including plant breeders, processing industries, seed distributors, agrochemical producers and marketers among others. Numerous private sector companies are integral components of the agricultural economy of many countries, where they play essential roles. During recent times, some companies in the private sector, particularly large multinationals have invested heavily in agricultural research, and particularly in biotechnology. The results of much of this research are not publicly available and cannot be used by national programmes to improve the situation of small farmers in developing countries. In the absence of regulatory frameworks in many developing countries, adapted crop varieties that incorporate elements of genetic engineering cannot be deployed. Moreover, private industry has concentrated heavily on a small number of crops of global importance. These are not necessarily the most important crops for small-scale farmers who often have to meet subsistence needs before they can consider cultivating crops to generate income.

The major issue in connection with the roles of the public and private sector is that they largely operate independently. In addition, the private sector is often able to attract the best qualified and most able individuals, to the detriment of the public sector. Public-private joint ventures, based on complementarity rather than competition, were seen by the working group as being vital to supporting agricultural production and plant breeding in developing countries. Co-operatives were considered a good example of how small-scale farmers could benefit from a more business-like approach to crop production and retail, especially in the seed sector. Encouraging small-scale agricultural enterprises was seen as a good way to boost development.

Donors and funding: With changing priorities among multilateral, bilateral and national donors, there has been a general move away from funding plant breeding. Not only do many donors have agendas that are not set according to recipient country needs, the agendas also change relatively rapidly in comparison with the long-term commitment that is necessary to realise benefits from plant breeding. A further marked failing is the fragmented way in which donors often operate. A coordinated, longer-term response was considered necessary to be able to achieve results in plant breeding that could benefit farmers and consumers in the developing world. Not only is it imperative that plant breeding programmes are funded over the long term, as plant breeders are not able to work without operational budgets, but capacity building also has to be addressed to make sure that there are trained personnel available to do the work.

Capacity building: Selected results from the questionnaire and consequent discussion indicated that declining plant breeding capacity, in terms of stagnant or reduced budgets and fewer released varieties, was confounded in some instances by increases in staff establishment and qualifications. One factor underlying this was changing age profiles of staff. Young, wellqualified people are attracted to other professions that offer better conditions and career prospects. Moreover, although biotechnology might feature among the more attractive subjects, it is very often isolated from practical plant breeding. What is required currently is a workforce that can use the tools of biotechnology in plant breeding. Without adequate numbers of suitably qualified staff, productive plant breeding will not be possible. Capacity building to achieve these ends was considered a priority area. The international agricultural research centers of the CGIAR have tremendous potential for contributing to capacity building. Not only do they conduct research on many of the world's most important crops, but they also represent an opportunity for NARS to send staff for training and retraining in a wide variety of subjects. Furthermore, ample field, laboratory and greenhouse facilities allow practical work to be pursued in cooperation with academic institutions elsewhere, including regional centers of excellence, which might provide training in plant breeding theory. Several CGIAR centres contribute substantially to NARS plant breeding efforts through supplying advanced germplasm, but they do not substitute for strong NARS: they strengthen NARS.

Regulatory issues: Plant breeders' rights (PBR) and intellectual property rights (IPR) were discussed as a means of encouraging private-public cooperation through promotion of public sector income generation, while protecting the interests of breeders. They are also able to accommodate participatory plant breeding. Subsistence farmers1, moreover, are exempt from much of the legislation involved in PBR, and are able to save and sow seed of protected varieties without the need for permission of the breeder. It was considered that PBR suited a variety of circumstances, although not all.

Prospects for regional cooperation in plant breeding were discussed. Where crops and cropping systems are common to a region it might be possible to mount a regional approach to plant breeding, and thereby reduce overall costs and limit unnecessary and unproductive duplication and competition. This will only be feasible under circumstances where associated legislation, particularly that regarding transfer of plant material across national borders, is harmonised. An added benefit of this approach might be that the stronger NARS could support weaker ones in a region.

Other forms of novel policy implementation to promote funding were discussed, including levies on products (not necessarily agricultural products) that could be fed back into the system to support research.

Biotechnology: Biotechnology is an integral feature of modern plant breeding research and practice. It is not however sufficiently well embedded in practical plant breeding in developing countries. It covers a wide range of methods, from relatively straightforward and cheap tissue culture, through marker assisted selection to genetic engineering. Some, and in a few cases all, of the biotechnologies associated with crop development have become part of plant breeding programmes in developing countries. Inadequate infrastructure, inadequate funding and lack of trained staff represent major obstacles to uptake however. These issues have to be addressed if plant breeding in developing countries is to be kept up-to-date. Capacity building in plant biotechnology will be most beneficial when staff are part of or closely linked with plant breeding programmes.

Next steps: It was proposed, and generally accepted, that there was a need for more consultations before any decisions could be reached on the way forward for strengthening national plant breeding and biotechnology capacity. It was suggested that a workshop should be organized with country representatives and decision-makers to increase awareness of the benefits of plant breeding and application of appropriate biotechnology to plant breeding. Presenting work on small-scale seed systems and establishing the links with plant breeding could follow this. Engagement with the private sector would be beneficial to establish common ground. Finally, having developed a strategy for strengthening national plant breeding capacities, meetings could be held with donors to discuss future funding. It was thought necessary to keep these steps separate and move relatively slowly rather than attempt to address all issues and include numerous stakeholders at a single meeting.

It was agreed that it would be useful to move forward with the national plant breeding survey, including more countries. There was a perceived need to address some of the issues concerning what has happened to plant breeding staff and what underlies the observed trends in staffing. It is essential to complete analysis of the data already received and publish it. Data up to 2004 have been requested and it was suggested that information on privatepublic linkages would represent a useful addition to those data already collected. The survey should be reviewed regularly by the participating countries.

The possibilities of setting up a special fund were discussed, with specific reference to linking with the private sector, and particulαrly for crops that do not currently interest the private sector. The IT was considered a possible mechanism to leverage funds to support small-scale seed enterprises with plant breeding activities. During the course of the discussion, it was suggested that FAO might consider publishing a text describing success stories in development of small-scale agro-industries, seed industries in particular, and the linkages with plant breeding.

Recommendations arising from the workshop

The following recommendations were formulated following three days of discussion by the working group on the major issues and problems associated with plant breeding and application of biotechnology in the public sector in developing nations. It was recognised that both rural and urban societies, and national economies of the developing world, are largely reliant on the agricultural sector. Nations with weak agricultural sectors, and particularly those without or with only poor plant breeding support, are at a disadvantage in comparison with those that are able to provide their farmers with improved crop varieties and appropriate technologies. Strengthening infrastructure and building capacity represent cornerstones of policies aimed at addressing current shortcomings. Strengthening sustainable agriculture through plant breeding and the application of appropriate biotechnology is of primary importance. The recommendations made by the working group were differentiated on the basis of major stakeholder groups.

Recommendations to FAO

1. Increase awareness among all stakeholders of the strategic importance of sustainable crop improvement through plant breeding.

2. Encourage all stakeholders in plant breeding in developing and developed countries to increase awareness of the value and impact of plant breeding and application of appropriate biotechnology on development.

3. Take note of the momentum that has been created through the Convention on Biological Diversity, the IT on PGRFA and the Global Crop Diversity Fund, and establish a platform for strategic thinking in a systematic manner for future development of sustainable use of PGRFA. This would necessarily include plant breeding and the application of biotechnology.

4. Consider establishing a fund to support sustainable plant breeding, for crops that have not received due attention of plant breeders and crop scientists, including genetic enhancement, in countries and for programmes requiring support. Capacity building would be a key component.

Recommendations to the Commission on Genetic Resources for Food and Agriculture (CGRFA) and the governing body of the International Treaty on PGRFA

1. Promote discussion of the use of PGRFA through plant breeding as an essential extension to conservation efforts. The IT might assist in strengthening the mechanisms and modalities for capacity building and technology transfer to developing countries (article 13 of the IT) to enable them to realise the potential of the PGR of their own farmers' varieties and the PGR made available through the multilateral system of the IT.

2. Encourage PGRFA users, particularly plant breeders, to contribute to CGRFA discussions and decision-making on sustainable use to complement efforts on conservation. Member countries of the CGRFA, where possible, should consider including specialists in sustainable use of PGRFA in their delegations to meetings of the Commission and IT Governing Body.

3. Give urgent consideration to capacity building and technology transfer to developing countries so that they are

able to use PGRFA in plant breeding programmes and contribute to the sustainable development of agriculture and allied activities.

4. Consider establishing a fund to support sustainable plant breeding for crops that have not received the due attention of plant breeders and crop scientists, including genetic enhancement, in countries and for programmes requiring support. Capacity building would be a key component.

Recommendations to NARS

1. Improve priority setting, incorporating awareness of markets and consumption chains and the paths of scientific and technological innovation.

2. Play an increasingly active role in decision making, especially in connection with capacity building and the need to strengthen it.

3. Adopt a multi-disciplinary approach to plant breeding, including biotechnology where appropriate.

4. Attempt to develop a solid and stable infrastructural and financial national base to ensure sustainability of crop improvement.

5. Seek to develop improved linkages, including those with farmers and the private sector, particularly with the informal seed sector where possible.

6. Increase awareness of the strategic importance of sustainable crop improvement.

7. Monitor and evaluate the impacts of IPR/IP on the development and effectiveness of breeding so that adjustments to strategies can be made on an informed basis.

Recommendations to Donors

1. Listen to all parties and promote awareness with a view to supporting plant breeding and its enhancement in developing countries.

2. Exert influence in developing national bases of support.

3. Improve coordination among donors.

4. Support networks to enhance capacity of national programmes, including through mechanisms such as the facilitating mechanism of the Global Plan of Action.

5. Support efforts to facilitate merging and integration of conventional and advanced breeding strategies.

6. Contribute to initiatives that foster sustained funding and long-term support needed to carry out plant breeding.

Recommendations to the CGIAR

1. Review survey documentation and advise on interpretation and strategic investments in crop improvement.

2. Balance donor demands and the needs of NARS and continue to support plant breeding through conducting research and capacity building according to the needs of national programmes and national programme staff.

3. Monitor the impact of capacity building efforts with a view to maintaining focused programmes.

4. Encourage and support cooperation at regional and sub-regional levels, and enhance south-south linkages.

5. Coordinate efforts to facilitate merging and integration of conventional and advanced breeding strategies.

6. Provide leadership in transfer of tools and technologies to NARS.

Recommendations to Policy-makers

1. Recognise the continued importance of agriculture to the national economies of developing nations and to the livelihoods of the rural and urban populations therein.

2. Recognise the contribution of plant breeding to improved food security and income generation in developing nations.

3. Understand the long-term nature of plant breeding and the needs for sustained support to ensure its success.

4. Encourage multi-stakeholder participation in decision-making processes that support conservation and use of PGRFA.

5. Encourage highly professional debate, relying on specialised representation, in forums such as the IT, on plant breeding and stress its importance in maintaining livelihoods in developing nations.

6. Harmonise policies related to sustainable use of PGRFA including seed systems.

7. Examine and understand the linkages between sustainable use and conservation of PGRFA.

8. Assist in implementing appropriate forms of rights for the developers and users of plant varieties used in agriculture in developing nations.

9. Monitor and evaluate the impacts of IPR/IP on the development and effectiveness of breeding so that adjustments to strategies can be made on an informed basis.

Agricultural Biotechnology

Agricultural biotechnology includes, but is not limited to:

- using our knowledge of cells to modify their activities.
- modifying cells or microorganisms using the techniques of recombinant DNA rather than traditional selective breeding.
- producing commercial quantities of desirable substances using microorganisms, cells or parts of cells.

Some examples of techniques used in agricultural biotechnology are:

- selective breeding that produces crops and livestock with specified characteristics.
- utilizing chemicals such as colchicine or radiation to induce mutations in genetic material.
- growing new plants using specific parts of another plant in a process called vegetative propagation. Grafting is one example that is important in fruit trees.
- encouraging the growth of a new plant in a prepared nutrient medium within a controlled environment.

Current examples of agricultural biotechnology products on the market are:

- corn plants having resistance to a particular herbicide used to kill weeds.
- corn plants producing a protein toxic to insect pests reducing the amount of insecticide use.
- tomatoes producing less of a hormone that causes tomatoes to ripen thereby extending the shelf life.
- corn hybrid plants growing in strongly alkaline soils that are found in the western US corn belt.
- cotton plants resisting cotton bollworms and tobacco budworms.

A large percentage of watermelons imported into the US annually come from Mexico with the remaining volume arriving from Central and South America. Imports have increased to meet the consumer demand for watermelon all-year-round.

Policy for Research in Biotechnology: A Model

The USDA portion of the "Proposal for a Coordinated Framework for Regulation of Biotechnology" (hereafter referred to as the December 3u, 1984 Notice) appeared at 49 FR 50897-50904. As a part of its policy perspective, USDA stated that agriculture and forestry products developed by biotechnology will not differ fundamentally from conventional products and that the existing regulatory framework is adequate to regulate biotechnology. USDA has both research and regulatory responsibilities for biotechnology activities. This document provides significant new information in both areas. Section II describes 1985 Federal Register notices concerning USDA policies and responsibilities for biotechnology. Included in this discussion is an explanation of the assignment of responsibilities within USDA for the oversight of USDA funded research and for the regulation of the products of biotechnology.

An understanding of the way in which USDA has divided these responsibilities should prove helpful to those in the private sector seeking review and/or approval of biotechnology applications. A new section III has been added describing USDA's policy for agricultural biotechnology research. USDA is publishing as a companion document, USDA Guidelines for Biotechnology Research that will closely parallel the NIH Guidelines. The USDA guidelines will be issued under the authority of the Food Security Act of 1985 (Pub. L. 99-198). This Act amended section 1404(2) of the National Agriculture Research, Extension, and Teaching Policy Act (NARETPA). The Amendment gave the Secretary of

Agriculture responsibility for establishing "appropriate controls with respect to the development and use of the application of biotechnology to agriculture." All USDA funded agriculture biotechnology research or research conducted at an entity receiving USDA funds would be subject to the USDA Guidelines for Biotechnology Research unless the specific research project is supported by and subject to the guidelines or regulations of another Federal agency.

These Guidelines would encompass all phases of agricultural biotechnology research, i.e. (1) Contained laboratory experiments; (2) specialized isolation research (e.g., greenhouse, biotron); and (3) environmental research release (e.g., controlled and segregated field plots). USDA hopes that entities not required to comply with the Guidelines would voluntarily adhere to the requirements. To encourage compliance, USDA proposes to adopt the NIH policy of providing the researchers not required to comply with these Guidelines the opportunity to have their new biotechnology research proposals reviewed by USDA. Those entities covered by the USDA Guidelines for Biotechnology Research would also be required to comply with any applicable statutes such as those set forth in section IV of this document, and any regulatory issues thereunder. The Secretary of Agriculture has established an Office of Agriculture Biotechnology (OAB), which will have primary responsibility for implementing and coordinating the Department's policies and procedures pertaining to all facets of biotechnology. This includes the conduct of laboratory and field research, exprimentation on biotechnology products prior to their commercialization, and all matters of oversight of biotechnology in agriculture. The new office will report to the Assistant Secretary for Science and Education through the authority provided in the amendment to the Food Security Act of 1985. The Assistant Secretary for Science and

Education will seek to establish an Agriculture Biotechnology Recombinant DNA Advisory Committee (ABRAC) and shall continue the responsibilities for agriculture formerly handled by the NIH-RAC during the last 10 years. The OAB shall operate in a close parallel manner to the Office of Recombinant DNA Activities (ORDA) of the National Institutes of Health. This includes the responsibility of the ABRAC and the implementation of the USDA Guidelines for Biotechnology Research. The NIH system is well respected both domestically and worldwide, and has achieved a high degree of efficiency in achieving broad confidence in the safety of new biological research conducted under its requirements. The OAB also will serve as a focal point for coordinating a National Biological Impact Assessment Program, which is to evaluate and monitor the potential impacts of biotechnological processes and products on safety and the environment.

Section IV contains USDA's regulatory policy statements for veterinary biological products, plants and plant products, meat and poultry products, and seeds. USDA stated in the December 31, 1984 Notice that while its existing regulatory framework is adequate, it would constantly reevaluate its regulatory position and should additional regulatory measures become necessary, amend its regulations (49 FR 50904). For veterinary biologicals regulated under the Virus-Serum-Toxin Act (VSTA), USDA has identified three categories which may be derived by recombinant DNA techniques or developed from hybridomas. The categories are based on biological characteristics and safety concerns, and are described fully in section IV(A). The first category consists of inactivated recombinant DNA-derived vaccines, bacterins, bacterin-toxoids, virus subunits, or bacterial subunits, as well as monoclonal products. This category presents no new or unusual safety or environmental concerns. The second category includes those products containing live microorganisms that have been modified by the addition or deletion of one or more genes. Such products will be evaluated under current regulatory policies and procedures to assure that the addition or deletion of specific genetic information does not impart increased virulence, pathogenicity, or survival advantages. The third category included products using live vectors to carry recombinant derived foreign genes for immunizing antigens and/or other immune stimulants. Characteristics of safety and transmission must be established fully before questions and concerns dealing with safety to humans, animals, and release into the environment can be answered and before such products can be considered for licensing. Section IV(A) also includes new information about revised USDA review procedures for the importation of cell cultures and hybridomas. A brief discussion is included about the proposed regulations implementing the provisions of the amendments to the VSTA contained in the Food Security Act of 1985. For organisms and products derived by the techniques of genetic engineering, USDA is proposing new rules to regulate organisms which are plant pests or which there is reason to believe are plant pests. It is USDA's policy to regulate certain genetically engineered organisms if the donor, vector/vector agent, or recipient organism is a member of a group of organisms that are known to contain plant pests, or if based on experience, USDA determines that a genetically engineered organism or product is a plant pest or if USDA has reason to believe that a genetically engineered organism or product is a plant pest. The proposed regulations are summarized in section IV(B). The USDA policy for regulating meat and poultry products and seeds derived through biotechnology remains substantially as stated in the December 31, 1984 Notice, and appears in section IV (C) and (D). A new section (V) has been added describing the scientific review mechanisms to be established by USDA to assit USDA Agencies in biotechnology research and regulatory decision-making. USDA has established a Committee on Biotechnology in Agriculture (CBA) chaired by the Assistant Secretary for Science and Education and the Assistant Secretary for Marketing and Inspection Services. A detailed summary of comments on the December 31, 1984 Notice and USDA responses appears as section VI. The comments are organized to conform to the form of the December 31, 1984 Notice, with general comments and responses on the USDA regulatory philosophy followed by comments and responses on specific aspects of USDA's regulatory structure.

Notices

Three Federal Register notices concerning the Department's biotechnology related activities have been published subsequent to publication of the December 31, 1984 Notice. On July 19, 1985, a document amending the delegations of authority of USDA to assign responsibility for these Research and regulatory activities (7 CFR Part 2) was published in the Federal Register (50 FR 29367-29368). In this document, the Secretary of Agriculture delegated responsibility to the Assistant Secretary for Marketing and Inspection Services to coordinate the development and carrying out of all matters and functions pertaining to the Department's regulation of biotechnology and to act as liaison on all matters and functions pertaining to the regulation to biotechnology between agencies within the Department and between the Department and government and private organizations. These responsibilities were further delegated from the Assistant Secretary for Marketing and Inspection Services to the Administrator of the Animal and Plant Health Inspection Service (APHIS). Also in this document, the Secretary of Agriculture delegated responsibility to the Assistant Secretary for Science and Education to coordinate the development and carrying out of all matters and functions pertaining to agricultural research involving biotechnology conducted or funded by the Department including the development and implementation of guidelines for oversight of research activities, and to act as liaison on all matters and functions pertaining to agricultural research in biotechnology between agencies within the Department and between the Department and other governmental, educational and private organizations. n1

n1 The Assistant Secretary for Science and Education oversees the research activities of the Agricultural Research Service (ARS), the Cooperative State Research Service (CSRS), the Extension Service (ES), and the Office of Grants and Program Systems (OGPS). The Assistant Secretary for Marketing and Inspecting Services oversees the regulatory activities of the Animal and Plant Health Inspection Service (APHIS), which includes Veterinary Services (VS) and Plant Protection and Quarantine (PPQ); the Agricultural Marketing Service (AMS); and the Food Safety and Inspection Service (FSIS). The policies and procedures of these agencies for biotechnology were described in the USDA portion of the coordinated policy statement at 49 FR 50899-50904. On September 23, 1985, USDA's APHIS published a notice which contained its policy statement and requirements for the control and protection of documents that contain confidential business information concerning biotechnology and the veterinary biologics program (50 FR 38561-38563). On November 14, 1985, the Office of Science and Technology Policy published a notice in the Federal Register announcing the establishment of the Biotechnology Science Coordinating Committee (BSCC) (50 FR 47174-47195). This Committee is to serve as an interagency forum for coordinating science issues related to research and commercial applications of biotechnology. The notice also stated that USDA will establish a Committee on Biotechnology in Agriculture (CBA) to assit in assuring that research and regulatory decisions and based on the best science available.

USDA Research Policy Statement

USDA supports research to promote and protect the general health and welfare of the people of the United States. n2 Research program include: Studies on production of food and agricultural processing and marketing; identity and development of new crop and animal sources of food, fiber, and energy; increased agricultural efficiency and reduction of dependence on petroleum-based products; development of improved management and conservation of soil, water, forest, and range resources. The programs are fulfilled through State, Federal, and private industry cooperative efforts.

n2 See Addendum for Research Legislative Authorities. In the areas of agricultural research relevant to biotechnology, many plant, animal, and microbial alterations have been developed for release through traditional genetic approaches such as mutagenesis and hybridization. In a complementary vein, beneficial introduction of organisms from abroad have established a sound base for research and regulatory oversight. The experience with these bases provide a substantial knowledge base for conducting evaluations of the safety and efficacy of biotechnology processes and products. USDA will evaluate the environmental impacts in the context of individual experiments that encompass the entire range of experimentation from contained facilities to open field testing. As knowledge and experience are gained, broadly applicable procedures and guidelines will be developed. Particular consideration will be given to the stability of engineered changes and the possibility that genetic elements might be transferred from one organism to another. Also important will be the development of data that will enable predictions of which organisms may become established in new ecosystems, and resulting environmental consequences. USDA considers products developed through biotechnological techniques as no different from those products resulting from research using conventional techniques providing appropriate research review is conducted with established protocols. Agricultural biotechnology research activities require appropriate review to avoid untoward effects on human health and the environment. USDA expects to rely on the existing network of scientific expertise in the agriculture research community. Thousands of plant selections, animal breeding lines, and microorganisms are tested annually at sites under varying climatic conditions through the Nation. This network of scientific expertise permits continual, open assessment of agricultural research and products of that research in the field. USDA has broad statutory authority to conduct and support research in wide ranging areas of agriculture. In addition to the authorities described in the matrix of Federal Laws related to biotechnology found in the Federal Register Notice of November 14, 1985 (50 FR 47174-47195) the Food Security Act of 1985 (Section 1404(2) of the National Agriculture Research, Extension, and Teaching Policy Act Amendments of 1985, Pub. L. No. 99-198), made the Secretary of Agriculture responsible for establishing "appropriate controls with respect to the development and use of the application of biotechnology to agriculture." Through this authority, and pursuant to the Delegation of Authority Pertaining to Biotechnology published in the Federal Register on July 19, 1985 (50 FR 29367-68), the Assistant Secretary for Science and Education will complete development of a national system of agricultural biotechnology research oversight in much the same manner that agriculture has been a part for the last 10 years through the NIH-RAC.

The Assistant Secretary for Science and Education has initiated the establishment of the Agriculture Biotechnology and Recombinant DNA Advisory Committee (ABRAC), to be managed through an Office of Agriculture Biotechnology (OAB) which is a parallel to the National Institutes of Health Recombinant DNA Advisory Committee (NIT-RAC) and Office of Recombinant DNA Activities (ORDA). The OAB will serve as the focal point for developing and coordinating USDA policies and activities pertaining to biotechnology research and will perform related interagency and public liaison functions. OAB will also assist in carrying out the responsibilities assigned to the Assistant Secretary for Science and Education, including the development and implementation of policies and procedures, and guidelines for the conduct of laboratory and field research. All federally-funded agriculture biotechnology research or research conducted at an entity receiving USDA funds will be subject to the USDA Guidelines for Biotechnology Research, which are published as a companion document to this policy statement, unless the specific research project is supported by and subject to the guidelines or regulations of another Federal agency. These Guidelines encompass the entire spectrum of degrees of containment in agricultural biotechnology research i.e.: (1) Contained laboratory experiments; (2) specialized isolation research (e.g., greenhouse, biotron); and (3) environmental research agricultural biotechnology release (e.g., controlled and segregated field plots). Research investigators not required to comply with USDA Guidelines will be encouraged to follow these Guidelines. To assure consistency, USDA adopted the model established by the NIH of providing such researchers with the opportunity to have their biotechnology research proposals reviewed as required by the Guidelines. The USDA Guidelines for Biotechnology Research require that research organization use the

Institutional Biosafety Committee (IBC) concept as established by NIH. This requirement assures that each research organization and its investigators employ a multidisciplinary team to assit in carrying out their responsibilities under the Guidelines. The IBC's, as described in the Guidelines, would consist of persons with relevant agricultural expertise in areas such as recombinant DNA technology, biological safety, physical containment, and ecology. Requests for review beyond IBC should be sent to the Office of Agriculture Biotechnology (OAB) through the Assistant Secretary of Science and Education, Room 324-A, Administration Bldg., Washington, D.C. 20250. These Guidelines also would require compliance with existing statutes of the USDA involving the movement of regulated organisms that require the issuance of a permit. The movement of microorganism injurious to plants and animals as well as the movement of certain non-indigenous plants and animals would continue to follow long-established procedures for USDA approval. After review, a permit, if needed, may be issued that allows movement. It is the responsibility of the research scientists to obtain that permit. The Assistant Secretary for Science and Education will complete establishment of a National Biological Impact Assessment Program (NBIAP) as indicated in the USDA Guidelines for Biotechnology Research. NBIAP would serve to assist USDA in the evaluation and monitoring of biotechnology research and impact over time. Coordination of NBIAP will be provided through OAB.

USDA Regulatory Policy Statements

The existing USDA regulatory authority for biotechnology was listed in the matrix of the December 31, 1984 Notice at 49 FR 50860-50874 and described in brief at 49 FR 50898-50899. The statutes considered most applicable to biotechnology applications are the Virus-Serum-Toxin Act

(VSTA) of 1913 (21 U.S.C. 151-158), the Federal Plant Pest Act (FPPA) of May 23, 1957 (7 U.S.C. 150aa-150jj), the Plant Quarantine Act (PQA) of August 20, 1912 (7 U.S.C. 151-164, 166, 167), the Organic Act of September 21, 1944 (7 U.S.C. 147a), the Federal Noxious Weed Act (FNWA) of 1974 (7 U.S.C. 2801 *et seq.*), the Federal Seed Act (FSA) (7 U.S.C. 551 *et seq.*), the Plant Variety Protection Act (PVPA) (7 U.S.C. 2321 *et seq.*), the Federal Meat Inspection Act (FMIA) (21 U.S.C. 601 *et seq.*), and the Poultry Products Inspection Act (PPIA) (21 U.S.C. 451 *et seq.*).

Veterinary Biological Products

Under the Virus-Serum-Toxin Act of 1913, 21 U.S.C. 151-158, the USDA exercises regulatory authority over all veterinary biologics imported into the United States or shipped or delivered for shipment interstate. Recent amendments contained in the Food Security Act of 1985 have extended this authority to products which are shipped intrastate or exported, and have given the Department additional enforcement mechanisms such as the power to detain and seize products. Under the VSTA, veterinary biologics may not be shipped or delivered for shipment if they are worthless, contaminated, dangerous, or harmful. Veterinary biological products must be prepared in a USDA-licensed establishment under regulations promulgated by the Secretary of Agriculture. Those products which are imported into the United States must be imported under a permit issued by the Secretary. The pertinent regulations for veterinary biologics are found in Title 9 of the Code of Federal Regulations, Parts 101 through 117. New regulations will be drafted to implement the provisions of the amendments to the VSTA. Such regulations will provide for a more comprehensive regulatory scheme, including seizure and condemnation and detention procedures. They also will establish procedures to be used in the issuance of special licenses and exemptions provided for by the legislative amendments. Veterinary biological products are defined in the governing regulations, 9 CFR 101.2(w) as "all viruses, serums, toxins, and analogous products of natural or synthetic origin, such as diagnostics, antitoxins, vaccines, live microorganisms, killed microorganisms, and the antigenic or immunizing components of microorganisms intended for use in the diagnosis, treatment, or prevention of diseases of animals." Licensing provisions for veterinary biological products and establishments are found in Part 102 of the USDA regulations (9 CFR Part 102). A product license requires the satisfactory completion of various requirements to assure purity, safety, potency, and efficacy of the products. The specific requirements were discussed in the December 31, 1984 Notice at 49 FR 50899. Pursuant to § 103.3 (a) through (g) of the USDA regulations, a person may be authorized to ship unlicensed biological products for the purpose of evaluating experimental products by treating limited numbers of domestic animals if USDA determines that the conditions under which the experiment is to be conducted are adequate to prevent spread of disease and approves the procedures set forth in the request for such authorization (9 CFR 103.3 (a)-(g)). Upon satisfactory completion of all requirements, including review and acceptance of labels, a U.S. Veterinary Biological Product License may be issued.

The application of new biotechnological procedures for the production of veterinary biological products is expanding constantly. For the purposes of licensing, biologics derived by recombinant DNA-techniques or developed from hybridomas, may be classified into three broad categories. This division is based upon the biological characteristics of the new products and the safety concerns they present, and is wholly analogous to the approach used in other veterinary biologics. The first category includes inactivated recombinant DNA-derived vaccines, bacterins, bacterin-toxoids, virus subunits, or bacterial subunits. These nonviable or killed products pose no risk to the environment and present no new or unusual safety concerns. Monoclonal antibody (hybridoma) products used prophylactically, therapeutically, or as components of diagnostic kits also are included in this category. The second category includes those products containing live microorganisms that have been modified by the addition or deletion of one or more genes. Deleted genes may code for virulence, oncogenicity, enzyme activity, or other biochemical functions. Added genes may result in the expression of new immunizing antigens or the production of novel biochemical byproducts such as beta-galactosidase. Precautions must be exercised to assure that this addition or deletion of specific genetic information does not impart increased virulence, pathogenicity, or survival advantages in these organisms which are greater than those found in natural or wild-type forms. Modifications also must not impart undesirable new or increased adherence or invasion factors, colonization properties, or intrahost survival factors. It is important that genes added or deleted do not compromise the safety characteristics of the organisms. In most cases it is expected that they will be improved, and would therefore not pose any new threat to humans, other animal species, or to the environment. The genetic information to be added or deleted must consist of well-characterized DNA segments. Required licensing data may include base pair analysis, sequence information, restriction endonuclease sites, as well as phenotypic characterization of the altered organism. A comparison is also required to be made between the genetically engineered organism and the wild-type form with respect to biochemical pathways, virulence traits, or other factors affecting pathogenicity. The third category includes products using live vectors to carry recombinant-derived foreign genes that code for immunizing antigens and/or other immune stimulants. Live vectors may carry multiple recombinant-derived foreign genes since they can carry large quantities of new genetic information. They also are efficient at infecting and immunizing target animal species. These properties, for example, make vaccinia virus recombinants very popular subjects for vaccine development programs.

Live vectors currently being evaluated by licensees, applicants, and other research organizations include vaccinia, bovine papilloma virus, adenoviruses, Simian Virus-40, and yeasts. Characteristics of safety and transmission must be examined before questions and concerns dealing with safety to humans, animals, and release into the environment can be answered and before such products can be considered for licensing. USDA will continue to avail itself of additional expertise from the Public Health Service "Interagency Group to Monitory Vaccine Development, Production, and Usage." This interagency committee will be utilized to consider potential human health hazards from the use of veterinary biological products and to review issues such as those arising from the potential effect of organisms potentially pathogenic to people or animals. Veterinary biological products prepared using modern biotechnological procedures such as recombinant DNA, chemical synthesis, or hybridoma technology will be treated similarly to products prepared by conventional techniques. The unlimited number and kind of products that may result from these modern biotechnology procedures make it impossible to define all requirements in specific terms. Each product is evaluated individually to determine what will be necessary to establish its purity, safety, potency, and efficacy. Scientific considerations may dictate generic areas of concerns or the use of certain tests for specific situations. Special assays, preferably using in vitro methods, may be required for potency and stability determinations. Additional tests may be required to assure safety, especially when live microorganisms are present in the biological products. USDA is authorized to issue three types of permits for importing biological products into the United States (9 CFR 104.2). A separate United States Veterinary Biological Product permit is required for each shipment of biological product to be imported. Permits are required for imported biological products used for research and evaluation, distribution and sale, or transit shipment only. Requests for application (U.S. Form 14-5) should be submitted to the Veterinary Biologics Staff, Veterinary Services, Animal and Plant Health Inspection Service, 6505 Belcrest Road, Hyattsville, Maryland 20782. To provide guidance to current or prospective manufacturers employing modern biotechnological methods, the following points are presented:

1. *Recombinant DNA-Derived Products.* Genetic information coding for a product of interest and other sequences not indigenous to the host are referred to as foreign DNA. Recombinant DNA technology encompasses the isolation, characterization, and expression of foreign DNA in organisms or vectors. The specific cloned nucleotide segment coding for the desired product or other foreign DNA segments must be defined in data supporting each license application. These data must also include a description of the source of the DNA and the nucleotide sequence. A vector is a cloning vehicle which provides a suitable origin of replication necessary for production of foreign DNA. Such replicons may be derived from plasmids, bacteriophages or viruses such as vaccina, bovine pipillomavirus, adenoviruses, or SV-40. Production of functional gene products depends on the efficient expression of cloned DNA-vector complexes in suitable host organisms. Tissue culture cells, bacteria,

yeasts, and virus cells may be used as hosts for replication of vectors. The mechanisms of transfer, the copy number, and the physical state of the constructed vector inside the host cell, integrated or extrachromosomal, must be described. USDA's licensing procedure for veterinary biological products derived from recombinant DNA involves a careful evaluation of each product on an individual basis to assure purity, safety, potency, and efficacy. Scientific and safety considerations may require specific safeguards and procedures in some situations. The USDA strongly recommends that all applicants establish Institutional Biosafety Committees which follow applicable provisions of the NIH Guidelines for Research Involving Recombinant DNA Molecules. USDA intends to propose guidelines which specifically relate to veterinary biological products. Amendments of the regulations and standards dealing with veterinary biologics will also be considered.

2. *Chemically Synthesized Antigens.* When the product consists of chemically synthesized polypetides, the appropriate amino acid sequences will mimic the antigenic site or epitope found in the native antigen where one exists. Procedures used to increase or prolong an immune response, such as coupling to carrier proteins or addition of adjuvants, must also be described. Immunological data derived from chemically synthesized peptides must be as definitive as those from natural antigens.

3. *Monoclonal Antibody Products.* The specificity and potency of monoclonal antibody will be compared with those of similar policyclonal antibody products where appropriate. The sensitivity and specificity of monoclonal antibody products used in diagnostic test kits and their potency characteristics when used therapeutically must be similar to conventional antibody. Monoclonal antibody must be derived from Master Cell Stocks which meet the applicable

requirements of 9 CFR 113.52. In addition, as is currently required, a description of cell cloning procedures, preparation, and characterization of cell passages must also be provided. The Outline of Production must describe all processes including scale-up, ascites fluid or cell culture supernatant preparation, purification, concentration, and inactivation, Mouse colonies must be screened to demonstrate freedom from adventitious agents, especially those detected by the mouse antibody production (MAP) test. If the MAP test discloses the presence of adventitious agents, the product shall not be released unless inactivation procedures approved by Veterinary Services have been performed and tests conducted to ensure proper application of the procedures.

4. *Master Seeds.* Bacterial or viral seed stocks used to prepare veterinary biological products must meet established procedures used to certify Master Seeds for biological products. The Master Seed for recombinant DNA-derived products may consist of a plasmid or virus carrying the inserted gene. This constructed plasmid is then introduced into the appropriate eukaryotic or prokaryotic expression system selected for vaccine production. Genomic DNA may also be transfected directly into a variety of mammalian cells. Alternatively, in such cases, the stable transfected cell could be considered as the Master Seed. The establishment of Master Seeds consisting of constructed plasmids or transfected cells requires submission of background information concerning the recombinant DNA procedures used to isolate, purify, and identify genetic material from one source and the modification used for inserting of this material into a new host. Data from cloning, isolation, proliferation, and selection of genetically unique cells would be retained by licensed applicants. In order to characterize adequately the foreign DNA used to code for a particular antigen, the manufacturer must provide

a nucleotide sequence analysis. Tissue culture-propagated cells from vertebrate animals used for vector propagation and antigen production must meet the requirements of 9 CFR 113.51 or 113.52. If a Master Seed has been accepted by Veterinary Services for use in a licensed product, further genetic modifications may be approved with reduced requirements for additional host animal efficacy studies. Each Outline of Production must be prepared in accordance with 9 CFR 114.9. Outlines must include procedures to ensure consistency in production and recovery of specific antigenic material. Recovery procedures must include removal of excessive antibiotic levels (9 CFR 114.10) and undesirable fermentation byproducts such as excessive levels of bacterial endotoxins. Serial release tests for purity, safety, and potency will be required. In addition product characterization tests may be required to demonstrate consistent gene expression.

Organisms and Vectors

Pursuant to the Act of February 2, 1903, (21 U.S.C. 111), and the VSTA, USDA has authority to issue such regulations and take such measures as may be deemed proper to prevent the introduction or dissemination into the United States for the contagion of any contagious, infectious, or communicable disease of animals and/or live poultry from a foreign country into the United States or from one State or territory of the United States or the District of Columbia to another. The importation into the United States or interstate shipment of organisms and vectors is regulated under 9 CFR Part 122. Organisms and vectors are defined in 9 CFR 122.1 as entities which may introduce or disseminate any cantagious or infectious disease of animals. Such substances may not be shipped interstate or imported without a permit. Permit applications must completely describe the substances, intended use, location of the permittee, and safeguards. A number of revised administrative and technical provisions have been instituted to expedite the USDA review and issuance of permits for importation or organisms and vectors which include cell cultures and hybridomas. No animal-origin biological materials, such as cell cultres, monoclonal antibodies, organisms, vectors, or related material, may be imported into the United States without a Veterinary Services (VS) Permit (VS Form 16-3A). To obtain a permit, an application (VS Form 16-3) should be submitted to: Import-Export Staff, Organisms and Vectors, VS, APHIS, USDA, 6505 Belcrest Road, Hyattsville, MD 20782. This is different from the permit required to import veterinary biologics pursuant to Part 104 of the USDA regulations governing such products (VS Form 14-5 and 14-6). Applicants must also complete the questionnaire entitled "Importation Information" and submit it with their application. Based upon the information submitted by the applicant, a determination will be made if the material to be imported requires safety testing to ensure it is free from livestock pathogens. Safety testing is conducted at the Foreign Animal Disease Diagnostic Laboratory (FADDL), Plum Island, New York. Applicants will be advised if a safety test is required and will be given an estimate of the cost for conducting the test. Applicants desiring to have material safety tested must enter into a Cooperative Trust Fund Agreement with APHIS, VS, and deposit in advance sufficient funds to cover the estimated cost. The Import-Export Animals and Products Staff will initiate the Cooperative Trust Fund Agreement. In order to expedite the procedure, VS may issue a permit for the material to be shipped to FADDL pending receipt of the funds and Cooperative Trust Fund Agreement. However, the signed Cooperative Agreement, plus the necessary funds, must be received by VS before testing can be scheduled at FADDL. Usually 60 to 90 days is needed for issuing a permit for importing material to Plum Island, New York, the completion of safety tests, and the transfer of the imported material to the applicant. A minimum of four vials, each containing at least 1 million cells from a uniform lot, is required for the safety testing. When the test is completed and a determination made that the imported material is free from livestock pathogens, the remainder of the imported material is released directly to the importer under conditions specified in the permit. If an importer wishes to import cell cultures and/or hybridoma cells on a regular basis, the applicant may enter into a continuous Cooperative Trust Fund Agreement with VS and establish an escrow account to ensure that unnecessary delays will not occur due to insufficient funds.

Each safety test utilizing susceptible host animals usually cost approximately $2,000 to $3,000. Sometimes it is possible to reduce the cost by pooling samples in one host animal test. Scientists at FADDL developed in vitro safety tests to detect certain livestock pathogens resulting in substantial cost savings for importers. The current cost of each in vitro test is approximately $500, depending upon the type of animal disease present in the country of origin as well as the intended use of the imported material. Safety testing may not be required for some cell cultures imported for human diagnostic purposes and research. Examples of material which could enter without safety testing include cultured human bone marrow cells, amniocentesis samples, and cells imported for karyotype analysis. Applications for such cell cultures will be considered individually. Permit applications are evaluated by a new classification scheme that correlates intended use of imported cell cultures with the level of safety testing conducted at FADDL. The following classification of cell cultures is based on intended use and generally indicates the level of safety testing required. Class

I Cell cultures to be used for the production of products such as vaccines, hormones, or other biologicals to be used in livestock, poultry, or for commercial distribution. Requirement: These cell cultures must be safety tested at FADDL using susceptible host animals, approved in vitro test, and/or laboratory animals. Class II Cell cultures to be used only for in vitro studies and not to be used in animals other than primates. Requirement: These cultures may not require safety testing. The material may be sent directly to the importer when no safety testing is required. The permit (VS Form 16-3A) will specify restrictions such as "FOR IN VITRO LABORATORY TESTS: DO NOT INOCULATE INTO LIVESTOCK, BIRDS, OR LABORATORY ANIMALS." Cell cultures imported under permit which do not require a safety test may not be distributed to other laboratories without prior approval from USDA, APHIS, VS. Applications for the distribution of imported material should be submitted to the USDA, APHIS, VS, Import-Export Staff, Organisms and Vectors. When appropriate, a review is conducted by the Administrator's Parent Committee on Organisms and Vectors. Members of this committee have wide expertise in evaluating safety. Clearance may also require testing in high security facilities at the Veterinary Services, FADDL, Plum Island, New York.

Plants and Plant Products

Pursuant to the authority granted by the Federal Plant Pest Act (FPPA) of May 23, 1957, as amended (7 U.S.C. 150 aa through 150 jj), and the Plant Quarantine Act (PQA) of August 20, 1912, as amended (7 U.S.C. 151 through 164, 166, and 167). USDA has regulatory authority over the movement into or within and through the United States of plants, plant products, plant pests, and any product or article which may contain a plant pest at the time of movement. These articles are regulated in order to prevent the introduction, spread, or establishment of plant pests new to or not widely prevalent in the United States. The regulations implementing this statutory authority are found in 7 CFR Parts 300 through 399. "Plant Pest," as defined by statute, means any living stage of any insects, mites, nematodes, slugs, snails, protozoa, or other intertebrate animals, bacteria, fungi, or parasitic plants or reproductive parts thereof, viruses, or any organisms similar to or allied with any of the foregoing, or any infectious substances, which can directly or indirectly injure or cause disease or damage in any plants or parts thereof, or any processed, manufactured, or other products of plants (7 U.S.C. 150aa(c)). "Movement," as defined by statute, means to ship, deposit for transmission in the mail, otherwise offer for shipment, offer for entry, import, receive for transportation, carry, or otherwise transport or move, or allow to be moved, by mail or otherwise (7 U.S.C. 150aa(g)). The current permit system requirements for the movement into or within and through the United States of plants, plant products, plant pests, and other articles regulated by FPPA and PQA were fully described in the December 31, 1984 Notice at 49 FR 50900-01.

The procedures for issuing permits for the movement of plant pests were discussed separately from plants, plant products and other articles which may contain plant pests at 49 FR 50901-02. USDA regulates the importation of noxious weeds through a permit system similar to that established for plant pests.

The existing regulations in 7 CFR Part 360 which designate plants as noxious weeds and establish procedures for obtaining an import permit were described at 49 FR 50902.

Regulation of the Introduction of Organisms and Products Altered or Produced Through Genetic Engineering Which are or Which There is Reason to Believe are Plant Pests

The FPPA and PQA are applicable to the movement of plants, plant products, and other articles and plant pests developed through genetic engineering if such plants, plant products, other articles, or plant pests present a risk of plant pest introduction, spread, or establishment. Under the authority granted by the FPPA and PQA, USDA is proposing new regulations which would impose restrictions on the introduction of organisms and products altered or produced through genetic engineering which are plant pests or which there is reason to believe are plant pests. In accordance with the provisions of the FPPA and PQA. USDA must determine the plant pest status of plants, plant products or articles to be moved into or within or through the United States. The evaluation process for determining what safeguards, if any, can be imposed which would allow the movement of the plant pest without risk that the plant pest would be disseminated were described in the December 31, 1984 Notice at 49 FR 50901-02. For genetically engineered material from dissimilar source organisms (inter-generic combinations), the determination may be complex. Information about genetically engineered organisms produced through the use of donor, vector/vector agent and recipient organisms that are from a list of known plant pests is needed in order that such organisms be properly regulated.

During the past year, USDA has received permit applications to move genetically engineered organisms into or through the United States. USDA is confident that organisms altered through genetic engineering will play a major role in increased plant yield and improved plant quality. However, a genetically engineered organism derived from organisms that are plant pests also presents a risk of plant pest introduction. The organisms themselves, the cultures in which they are transported, or their packaging may be contaminated with plant pathogens. Genetic alteration may create a plant pest new to and not widespread in the United States. It is necessary, therefore, to establish appropriate safeguards to prevent the introduction of genetically engineered organisms that pose a threat to agriculture.

Other genetically engineered organisms that are not plant pests or where there is no reason to believe such organisms are plant pests would not be regulated. New data have to be required in order to properly evaluate permit applications for those organisms which are plant pests or which there is reason to believe are plant pests. A determination was made that additional data requirements would be incorporated into proposed regulations for those genetically engineered organisms which are of concern under the provisions of the FPPA and PQA. USDA is publishing as a companion document in the "proposed rules section" of this issue of the Federal Register its proposed regulations pertaining to organisms and products altered or produced through genetic engineering which are on plant pests or which there is reason to believe are plant pests.

The proposed regulations would establish a new part entitled, "Introduction of Organisms and Products Altered or Produced Through Genetic Engineering Which Are Plant Pests or Which There is Reason to Believe are Plant Pests", in Title 7 of the Code of Regulations (7 CFR), pursuant to the authority of the FFPA, as amended (7 U.S.C. 150aa-150jj) and the PQA, as amended, (7 U.S.C. 151-164, 166, 167). Such proposed regulations would regulate the importation into and movement within and through the United States as well as prevent the release into the environment of certain organisms, or products altered or produced through genetic engineering, which are plant pests or which there is reason to believe are plant pests.

The proposed regulations would restrict the "introduction" of certain organisms and products altered or produced through genetic engineering, referred to as "regulate articles." In this context, "introduction" means to move into the United States, to release into the environment, or to move interstate, or any attempt thereat." "Release into the environment" means "use of a regulated article outside the constraints of physical confinement that are found in a laboratory, contained greenhouse, or fermenter or other contained structure." USDA's proposed regulations, which are designed to prevent the release into the environment of genetically engineered organisms which are plant pests or which there is reason to believe are plant pests are consistent with the legislative intent of the FPPA.

The FPPA was enacted in 1957 and was intended as "gap filing" legislation for the purpose of protecting American agriculture against invasion by plant pests and diseases which are new to or not theretofore known to be widely prevalent or distributed within and throughout the United States. The FPPA also provides USDA with authority to regulate insects or pests that might later be found to be injurious to cultivated crops. The release into the environment of a genetically engineered plant pest is tantamount to the introduction of a plant pest which is new to and not theretofore known to be widely prevalent within and throughout the United States and subject to regulation under the FPPA. It should be noted that "regulated article" would be defined as any organism or product altered or produced through genetic engineering, if the donor organism, recipient organism, or vector or vector agent belongs to a group of organisms designated by the proposed regulations as having plant pests or any organism or product which

USDA determines is a plant pest or which there is reason to believe is a plant pest. Under USDA's proposed definition, certain microorganisms would be excluded if the recipient microorganism is non-pathogenic, is non-infectious. and otherwise not a plant pest, and resulted from the addition of genetic material that is well characterized and contains only non-coding regulatory regions. Restrictions would be required for regulated articles because they are plant pests, or because USDA has reason to believe they are plant pests. The proposed regulations would require that a person obtain a permit prior to the introduction of a regulated article and would list specific conditions required for the introduction of a regulated article. The regulated article could be introduced only if all conditions in the proposed regulations as well as all conditions specified on the permit were met.

It is important to note that in considering whether a permit can be issued for the introduction of a genetically engineered organism, USDA will perform the same comprehensive analysis that is used in determining whether a permit can be issued for the movement of a "conventional" plant pest. Such asessment shall include an examination of the factors that were discussed in the December 31, 1984, Notice at 49 FR 50901-02 as part of the evaluation process for determining what safeguards can be imposed which would allow the movement of a plant pest without risk of dissemination. These factors are oriented toward an examination of the ecological and environmental effects of a release of the genetically engineered organism or product into the environment. The proposed regulations also contain provisions for a certificate of exemption for those organisms or products altered or produced through genetic engineering that are not subject to the proposed regulations. A person seeking to introduce an exempt article could voluntarily request a certificate of exemption to facilitate the introduction of the organism or product. The proposed regulations provide a list of groups of organisms which are plant pests or contain plant pests. If the donor, vector/vector agent, or recipient of the genetically engineered organism is derived from an organism on the list of organisms containing plant pests, such genetically engineered organism would be deemed a "regulated article".

As defined in the proposed regulations, a plant pest includes microorganisms such as bacteria and viruses, and thus a "regulated article" may be a microorganism unless it meets the provisions for exclusion. It is important to note that in some instances certain microorganisms will be subject to joint regulation by USDA and EPA. USDA has jurisdication over certain microorganisms under the FPPA and PQA if the mircorganisms are a plant pest. EPA would have jurisdiction under the Toxic Substances Control Act (TSCA) if the microorganism is deemed to be a "new" microorganism or under the Federal Insecticide, Fungicide, and Rodenticide Act, as amended (FIFRA) if the microorginism is to be used as a pesticide. Because each Agency has a different statutory mandate, certain jurisdictional overlaps cannot be avoided. However, EPA and USDA will work cooperatively and simultaneously in the evaluation of genetically engineered microorganisms that fall under the jurisdication of both Agencies. To expedite the review of these microorganisms each Agency will appoint contact persons to coordinate the review to ensure data requests are not duplicated. The specifics of which microorganisms will be subject to dual Agency review, or primarily single Agency review, is set forth in the preamble of USDA's proposed regulations being published as a companion document to this policy statement. That document should be consulted for further information. A key to determining whether a genetically engineered organism will be regulated by USDA is the list of organisms containing plant pests that appears in § 340.2 in proposed Part 340. USDA acknowledges that this is not an exhaustive list, and that it does not attempt to list every pest species. Comments are welcome on the list as well as on other parts of the proposed regulations. In order to solicit as many comments as possible on the list and all other parts of the proposed regulations, USDA has scheduled public hearings in Washington, DC and Sacramento, California, during the 60-day comment period. The time and place of the public hearings as well as the address to send written comments is specified in the preamble to the proposed regulations. USDA believes that through the submission of detailed comments and full participation by public and private interests, USDA will be able to promulgate a final regulation that will prevent the introduction and dissemination of genetically engineered organisms which are plant pests or which there is reason to believe are plant pests, yet not impede the development of biotechnology.

Meat and Poultry Products

The Food and Safety Inspection Service (FSIS) is responsible for assuring the safety, wholesomeness, and proper labeling of food products prepared from domestic livestock and poultry. The Federal Meat Inspection Act (FMIA) and the Poultry Products Inspection Act (PPIA) require FSIS to inspect cattle, sheep, swine, goats, equines, poultry, and food products prepared from them which are intended for use as human food to assure that they are wholesome, not adulterated, and properly labeled, marked, and packaged. Inspection under these statutes is mandatory. The cost of inspection, except for overtime and holiday inspection work, is required to be borne by the USDA. Food, animals and animal products, other than those required to be inspected under the FMIA and PPIA, may be inspected under a voluntary, reimbursable inspection program established under the Agricultural Marketing Act of 1946. Within the framework of food safety statutes, FSIS has developed regulations for research on animals that are administered experimental animal drugs, biologics, and pesticides (9 CFR 309.17 and 381.75). These regulations state that no animal used in any research investigation involving an experimental biological product, drug, or chemical shall be eligible for slaughter at an official establishment unless certain conditions are met. These conditions include any of several different ways of demonstrating that the use of such biological product, drug, or chemical will not result in the products of such animals being adulterated. *Products Subject to Review.* FSIS anticipates that many food animals which are subject to the new techniques of modern biotechnology will not differ substantially in appearance, behavior, or general health from currently inspected cattle, sheep, swine, goats, equines, and poultry. They would be subject to the same inspection procedures and regulations as tradionally inspected food animals. FSIS is aware that some genetically engineered animals, such as mosaics, chimeras, and some hybrids, may differ substantially from animals that are inspected currently under the FMIA and PPIA. If such animals are ever intended for use as human food and are presented for inspection at an official establishment, a decision would have to be made as to whether such animals were covered under the FMIA or PPIA, and if not, whether the FMIA and PPIA should be amended to require inspection of such animals and their products.

Implementation of Review Authority. FSIS's approach toward the review of food animals resulting from the techniques of modern biotechnology consists, in general, of two phases. The first, an experimental phase, focuses on the experimental aspects of vector administrative, gene transfer and gene expression. Since artificial vectors used in animal gene transfer may be considered as either animals drugs or animal biologics, their administration to food animals would be covered under the current regulations on animals used for research (9 CFR 309.17 and 381.75). The requirement that an animal carcass intended for use as human food not be adulterated may require that certain phenotypic, biochemical, and microbiological parameters not be exceeded before the animal can be slaughtered for human food. Depending on future developments, FSIS may amend the regulations (9 CFR 309.17 and 381.75) to provide further assurance that the products of animals genetically engineered by certain techniques are not adulterated. The second phase would be carried out under existing regulations (9 CFR Parts 301 through 381) and would focus on the commercial development, production, inspection and labeling of food animals and food animal products.

Seeds

The Federal Seed Act (FSA) (7 U.S.C. 1551 *et seq.*) defines USDA regulatory authority over the importation and interstate shipment of agricultural and vegetable seeds. It does not apply to the production or intrastate distribution of seeds or to seeds other than agricultural or vegetable seeds ("agricultrual seeds" are grass, forage, and field crop seeds). The FSA prohibits interstate shipment of seed that contains noxious weed seeds at levels in violation of the laws of the State of destination or in excess of levels allowed by the Secretary of Agriculture. This provision applies primarily to seed adulterated with noxious weed seed. In a few instances, however, States have determined that a particular variety of agricultural or vegetable seed is itself a noxious weed. In these instances, FSA prohibits the interstate shipment of the seed into those States. The FSA also allows the Secretary to prohibit the importation of agricultural and vegetable seed which is adulterated with noxious weed seed or which is unfit for seeding purposes. The authority granted to the Secretary by the FSA to prohibit the interstate shipment or importation of seeds which are found to be detrimental to the agricutural interests of the United States applies to seeds genetically engineered with the modern biotechnology to the same extent as any other seeds.

Scientific Review Mechanisms

The manner in which both regulation and oversight of research in agriculture-related biotechnology evolves and is implemented in the United States will have a direct impact on the competitiveness of U.S. industry in both domestic and world markets. Inconsistent or unnecessary procedures for regulation and research will place the U.S. scientific effort and U.S. producers at a substantial disadvantage. It also is important that safeguards be built into biotechnological research processes, and that releases be based on careful evaluations while further experience is being gained. Therefore, USDA feels that such regulatory and research decisions must be based on the best science available.

While the responsibilities within USDA for biotechnology reside with the Assistant Secretary for Science and Education and the Assistant Secretary for Marketing and Inspection Services as the delegates of the Secretary of Agriculture, in carrying out their respective responsibilities based on the best science available, they would be able to take advantage of the expertise and perspectives within the Federal Government through a committee to be called the Committee on Biotechnology in Agriculture (CBA). The CBA, to be chaired by these two Assistant Secretaries, will function both as a policy body in the USDA and a bridge between its research and regulating structures.

Committee on Biotechnology in Agriculture

The objectives of the CBA will include:

To provide advice, when requested, on initiatives, proposals, and policy for agriculture-related regulation and research, and assist in the coordination of these activities; To review scientific issues submitted by agencies within the Department; To assist in identifying data gaps for basic research in agricultural biotechnology; To foster public awareness of the scientific issues in biotechnology; To provide Departmental support for participation in the FCCSET BSCC. USDA expects that the CBA also will utilize existing cooperative entities (e.g., other Federal agencies, universities, State regulatory officials, the public sector, and industry) to acquire, when necessary, information for addressing those issues submitted to it. Such entities, when requested, can provide technical support for sound regulatory and research decisions regarding the use of biotechnology in agriculture and foresty. These entities offer a vast scientific resource upon which USDA can draw.

Summary of Comments

USDA received the comments of one hundred-two (102) respondents, one-half of whom commented specifically on the USDA policy statement. Although USDA agencies considered all comments on the coordinated policy proposal, this response is confined to comments on the USDA portion of the notice. The two largest categories of respondents were business and academic, followed closely by associations representing these interests. Comments came in lesser numbers from environmental and public interest groups, individuals, law firms, and foreign governments, as well as the National Institutes of Health Recombinant DNA Advisory Committee (NIH-RAC) and a member of the U.S. Congress The USDA response to the comments follows the form of the original notice, with a discussion of comments on regulatory philosophy followed by a response to comments on the regulatory framework.

Comments on the Nature of Products of Modern Biotechnology: Fourteen respondents stressed their agreement with the USDA statement that "agriculture and forestry products developed by modern biotechnology will not differ fundamentally from conventional products," while six commenters dissented. Three respondents felt that genetic engineering across species barriers did create a potentially different product and the possiblity of unique ecological effects. Concern about the "need for public trust" and public assurance on safety and ethical issues was stressed by three commenters. Seven respondents agreed with USDA that "to date, no unique or safety problems have been associated with products of genetic engineering," but four of the same commenters who view biotechnology products as fundamentally different from conventional products stressed that the potential exists for safety problems with biotechnology applications.

Response: USDA recognizes the importance of ecological effects and the need for developing procedures responsive to public concerns about safety. Although USDA's regulatory philosophy remains as stated, additions to regulatory procedures are being proposed for genetically engineered plants and plant products and veterinary biologics produced by biotechnology (see section IV). The previously discussed delegations of authority within USDA for biotechnology increase the effectiveness of the administration of current and proposed regulatory procedures affecting the products of modern biotechnology. For veterinary biological products, USDA is currently developing additional procedures pursuant to the VSTA, as amended, for evaluating requests to conduct experimental field trials with live vectors containing genetically engineered organisms or to support product license applications. The procedures being developed consider the parental organism and the effect of the gene alteration on the genetic properties of the recipient, especially the survival, reproduction, and dispersal characteristics. A careful analysis of the genetics, biology, and ecology of the wild-type and modified microorganisms will provide as reasonable prediction of the risks which might be associated with use of the altered organisms. USDA is proposing regulations pursuant to the Federal Plant Pest Act (FPPA) and the Plant Quarantine Act (PQA) for regulating the introduction of certain organisms of products thereof altered or produced through biotechnology which are plant pests or may become plant pests. This proposed rule should assist USDA in assessing the ecological effects of the release of such genetically engineered organisms into the environment. Guidelines for oversight of agricultural biotechnology research funded by USDA will be issued under the authority of the Food Security Act of 1985. USDA also is establishing scientific review mechanisms to assist in research and regulatory decisions. These proposed modifications in the procedural framework are described as a part of the final policy statement for veterinary biologics, plants and plant products, research, and scientific review mechanisms.

Comments of the Adequacy of Existing Authority: Thirteen commenters agreed with USDA that its existing regulatory framework is adequate for biotechnology applications, and nine favor the case-by-case approach under existing authority. Five commenters felt that new legislation is or may be needed; two of the five oppose the case-by-case approach.

Response: USDA has examined its statutory authority for regulating biotechnology products and processes, and USDA agencies have processed licensing and permit applications under the existing statutes. The existing authority is considered adequate at this time. Established procedures, with the proposed modifications, can be adapted effectively to handle biotechnology applications. USDA is currently considering genetic engineering applications on a case-by-case basis using existing authority.

Comments on Need for Procedures and Guidelines: Sixteen respondents commented that USDA had not outlined procedures for the review and approval of genetically engineered products. Twelve respondents stressed the need for flexibility, and six requested sunset provisions in USDA biotechnology regulations.

Response: The USDA policy statement of December 31, 1984, did outline procedures currently used for the review and approval of certain genetically engineered products. In considering license applications for genetically engineered veterinary biologics, USDA follows the standards and procedures applicable to all such products found in § § 101-117 of the applicable regulations and standards (9 CFR 101-117). In the December 31, 1984 Notice, USDA offered supplementary guidelines for licensing such products. New procedures are being developed to evaluate production and testing of veterinary biologics derived through use of genetic engineering techniques. The information needed for proper evaluation will depend on the parent organism and the effect of the gene alteration on the genetic properties of the recipient. A paper describing the USDA licensing policy for biologics produced by recombinant DNA technology was presented at the Joint International Association of Biological Standardization/World Health Organization Symposium on "Standardization and Control of Biologics Produced by

Recombinant DNA Technology," Geneva, Switzerland, 1983 (published in *Developments in Biological Standardization,* V. 59, pp. 167-173, S. Korgel, Basel, 1985). The paper describes requirements for plasmid/vector characterization and stability, and correlation to conventional Master Seed concepts, as well as methodology which can be used to monitor antigenic expression, concentration, purification, and stability testing during production and recovery.

The movement of genetically engineered products which are plant pests and present a risk of plant pest introduction or spread is regulated by 7 CFR 330.200 implemented pursuant to the FPPA and PQA. The movement of organisms and vectors which may cause desease in animals is regulated under 9 CFR Part 122. USDA realized that the statement left unanswered some questions about the means for review and approval of various genetically engineered products. The proposed regulations described in section IV(B), implemented under the authority of the FPPA establish permit requirements for the "introduction" of organisms altered or produced by genetic engineering which are or may become plant pests. The regulations would be flexible because organisms determined not to be plant pests would be exempt, and this category could be expanded in the future to include organisms whose plant pest status is currently uncertain and therefore restricted. It is hoped that the discussion in section IV(B) of this policy statement answers any remaining questions about the review and approval procedures for such genetically engineered products.

Comments on Confidential Business Information (CBI): Six commenters representing business and scientific interests expressed concern about the protection of "confidential business information" in the USDA regulatory process while two public interest groups stressed the "public's right to know."

Response: The USDA regulations implementing the Freedom of Information Act (FOIA) (5 U.S.C. 552) are found in 7 CFR 1.1-1.16. The FOIA provides that Federal agencies must make available to the public all records not specifically exempt from disclosure. Exemptions include "trade secrets and commercial or financial information." (5 U.S.C. 552(b)(4)). On September 23, 1985, USDA's APHIS issued a policy statement on the protection of privileged or confidential information (50 FR 38561-38563). This policy statement establishes requirements for the control and protection of documents received by APHIS that contain privileged or confidential business information concerning biotechnology and the veterinary biologics program. The procedures established conform to the FOIA requirements for both protection and disclosure.

Comments on Use of NIH Guidelines: Four respondents questioned the USDA requirements that manufacturers of veterinary biological products using recombinant DNA technology follow the National Institutes for Health Guidelines for Research Involving Recombinant DNA Molecules (NIH Guidelines). One respondent thought USDA implied that all people "who work with recombinant DNA plants" would be required to comply with the NIH Guidelines, and requested procedural guidelines for industry.

Response: The USDA does not require that the manufacturers of veterinary biological products or plant products of recombinant DNA technology follow the NIH Guidelines. However, USDA strongly recommends that all license applicants for veterinary biologics follow appropriate provisions of the NIH Guidelines, such as those regarding the establishment of an institutional biosafety committee.

USDA intends to propose guidelines that will parallel closely the NIH Guidelines, and it intends to recommend strongly that entities not required to follow the USDA guidelines do so voluntarily.

Comments on Importation of Cell-Lines: Three associations representing biotechnology companies requested that USDA take steps to reduce delays in the clearance and testing procedures required for the importation of biotechnology-derived products and cell-lines. On February 12, 1985, the Association of Biotechnology Companies (ABC) delivered a report on USDA importation guarantine issues to the APHIS Parent Committee for Foreign Pathogens and Vectors. This report was an attachment to the ABC comment letter.

Response: The USDA has instituted a number of revised administrative and technical provisions to expedite the issuance of permits for importation of organisms and vectors which include cell cultures and hybridomas. A supplementary questionnaire, designed to insure adequate information on cell cultures and products from recombinant DNA or hybridoma technologies, now accompanies each permit application. Applicants are advised whether or not a safety test is required and a cost estimate is given. Safety testing may be conducted concurrently with the administrative review of the permit application, but only at APHIS' Foreign Animal Disease Diagnostic Laboratory (FADDL) at Plum Island, New York. New test procedures have reduced the cost of safety testing, and the cost per sample can be further reduced by conducting a safety test with pooled samples. Permit applications are evaluated using a new classification scheme that equates intended use of imported cell cultures with the level of safety testing required at FADDL. Class I cell cultures, employed in the preparation of products such as enzymes, vaccines, or hormones for commercial use, are subject to complete safety testing. Class II cell cultures, used only for in vitro studies and not to be used in animals other than primates, are subject to a lesser degree of testing.

Comments on Risk Analysis: Seven respondents discussed the issue of risk assessment or risk/benefit analysis of biotechnology applications. Comments varied from a recommendation that "standard risk assessment methodologies" be adopted by all agencies to a warning against attempting to regulate the "hypothetical and imaginary "potential' dangers" of recombinant DNA techniques.

Response: The National Environmental Policy Act (NEPA) applies to USDA actions. The "APHIS Guidelines Concerning Implementation of NEPA Procedures" (44 FR 50381, August 28, 1979) would be used to make an environmental assessment or environmental impact statement of the effects of a proposed release of a genetically engineered organism regulated pursuant to the VSTA, the FPPA and PQA, and related statutes. A formal risk management procedure based on a wide variety of safety concepts will be used to evaluate systematically proposed releases. The information required by any new regulations promulgated under the FPPA and PQA would be used to prepare the environmental assessment for release of a genetically engineered product which is a plant pest or may become a plant pest.

In normal husbandry and laboratory practices, veterinary biological products normally are not considered to be released into the environment. In the event that a conventionally prepared or recombinant derived product would be considered to be released into the environment, the issuance of a license or import permit would require compliance with procedures being developed and interagency approval. The procedures under development consider the parental organism and the effect on the gene alteration on the genetic properties of the recipient, especially the survival, reproduction and dispersal characteristics. Safety, ethics, and policy issues in agricultural biotechnology research will be overseen by the Committee on Biotechnology in Agriculture (CBA) and such supporting technical advisory groups as may be established by the USDA agencies. Currently, all USDA and USDA-sponsored research involving recombinant DNA must be cleared prior to initiation for compliance with the NIH Guidelines.

Comments on Jurisdiction: The potential for overlapping jurisdiction in the policy notice drew the largest number of comments. Eighteen respondents pointed out that both USDA and EPA propose to regulate agricultural microorganisms. Respondents representing the interests of the veterinary biologics industry contended that a jurisdictional dispute between USDA and FDA delayed the approval of bovine interferon. While generally supporting the concept of the memorandum of understanding (MOU) between USDA and FDA to resolve jurisdictional disputes, one respondent challenged the legality of the MOU, noting that it contains the statement that "animal biological products generally act through a specific immune process," while USDA's current regulations do not restrict its jurisdiction to products operating through such a mechanism of action. Industry respondents also pointed out that the intrastate producer of veterinary biologics is not regulated by USDA. Two firms and one industry association urged prompt Federal oversight action so that States do not act independently to regulate biotechnology products.

Response: USDA agrees that there is the potential for overlapping jurisdiction among the Federal agencies involved in regulating biotechnology products. USDA and EPA representatives have discussed jurisdiction over genetic engineering applications since 1983. USDA and EPA have begun to establish a regulatory procedure for reviewing certain submissions of genetically engineered microorganism applications, a procedure which has resulted in joint consultation on several proposals for release into the environment of organisms altered by genetic engineering.

For veterinary biologics regulated under the VSTA, use of procedures currently under development will increase USDA effectiveness in evaluating biotechnology license and product applications. The MOU between USDA and FDA was published on June 8, 1982, in an attempt to resolve the issue of new products which fall into the questionable definitional area between animal drugs regulated by FDA and animal biologics regulated by USDA. An interpretation by some that the term animal biologics only includes substances that act through a specific immune process has resulted in some confusion. There is nothing in USDA's current regulations or law which restricts its jurisdiction to products acting solely through this mechanism of action, and because of this fact, the memorandum qualifies its reference to specific immune process by the word "generally." Although efforts will be made to clarify the issue further, it should be noted that there appears to be little uncertainty about whether a particular product is a veterinary drug or biologic. The Food Security Act of 1985 contains amendments to the VSTA that extend USDA's jurisdiction to veterinary biologics which are shipped intrastate or exported. The provisions of the amendments are discussed more fully in Section IV.

Comments on the National Biological Impact Assessment Program (NBIAP): Seven respondents commented on the NBIAP, the proposal by the National Association of State Universities and Land Grant Colleges (NASULGC) for establishing a program to assess genetically engineered organisms before they are released into the environment. Three commenters — a member of Congress, a spokesperson for a biotechnology firm, and an officer of an environmental organization — posed questions about the proposal. The questions concerned the NBIAP's statutory or regulatory status; its relation to other USDA agency operations and other Federal agency operations; the processes of risk assessment to be used; its adequacy to review an increasing volume of products; and the appropriateness of biohazard committees as vehicles for review of commercial processes and products. Four respondents representing NASULGC institutions endorsed the proposal stating the view that the agricultural research community has the capability to develop guidelines and assess impacts of biotechnology research and commercial products. The major goal of the program was thought to be insuring the safety of society and the environment.

Response: NBIAP is a scientific advisory system that would be available to the Assistant Secretary for Science and Education. By this system the USDA can draw upon the best experience available from scientists in universities, Federal laboratories, and industry to help assess the risks involved in the processes and products from RDNA work in biotechnology. NBIAP shall act in an advisory capacity and is in no direct way a part of the formal approval process. It is available to provide assessment, but is not a mandatory process.

Comments on Definitions, Terms, and Data Requirements: Five respondents recommended changes in the definitions, terms, data requirements or classification used by USDA in the notice. Each recommendation is discussed below. Two respondents commented on the USDA statement of licensing policy for veterinary biologics produced by modern biotechnical methods at 49 FR 50899-50900. Under the heading "1. Recombinant DNA-Derived Products," a manufacturer of veterinary biologics questioned the need to provide the entire nucleotide sequence of a foreign DNA being cloned into a vector.

It is USDA's position that in order to characterize adequately the foreign DNA used to code a particular antigen, the manufacturer should provide a nucleotide sequence analysis. The construction of the vector used for expression of the cloned nucleotide sequence also should include source and function of the component parts of the vector, i.e., origin of replication, antibiotic resistance genes, promotor, enhancers, etc. The manufacturer also questioned the data requirement under the heading "2. Chemically Synthesized Antigens" concerning the persistence of the immune response following administration of the synthetic peptide.

The USDA feels that a major concern with the use of synthetic peptides is the development persistence of the immune response. USDA does not intend to require more stringent efficacy data than that necessary to support a veterinary biologic license application employing natural antigens. However, immunological data derived from chemically synthesized peptides must be as definitive as the serological response from natural or nonsynthetic antigens. With respect to the next sentence in the policy statement, an individual respondent proposed a change from the term "antibody response" to "immune response." It is true that the term used in the sentence "Procedures used to increase or prolong an antibody response . . ." is somewhat limiting and can create confusion between B-cell and T-cell response.

Therefore, the recommendation to replace "antibody response" with the term "immune response" with the term "immune response" is accepted, since both T-cell responses as well as T-cell/B-cell interactions would be included in the statement. On the subject of plants and plant pests, a plant pathologist commented on the references to *Pseudomonas syringae* as plant pathogens under the heading "ice nucleation negative bacteria" at 49 FR 50902. The respondent noted that none of the strains of *Pseudomonas syringae* currently proposed for use are plant pathogens and that it would be more correct to call *P. syringae* plant-associated bacteria, some of which are pathogens. USDA will clarify future references to these organisms as the respondent suggests. According to current practice, and under the proposed FPPA regulations, an applicant for a USDA permit to import or move *Pseudomonas syringae* would be required to submit data to show whether of not the strain was a plant pest.

Addendum — Research Legislative Authorities

The USDA is authorized under its Organic Act (7 U.S.C. 2201 et seq.) and other legislation to conduct and support research in wide ranging areas of agriculture. Examples of such other laws include: The Alcohol Fuels Research (7 U.S.C. 3154); the National Latex Commercialization and and Economic Development Act (7 U.S.C. 178-178n); the Animal Health and Disease Research Act (7 U.S.C. 3195); Special Research Grants (7 U.S.C. 450i(c)); The National Aquaculture Act (16 U.S.C. 2801 et seq.); the Cotton Research and Promotion Act (7 U.S.C. 2101 et seq.); the Potato Research Information Act (7 U.S.C. 2611-2627); the Egg Research and Consumer Information Act (7 U.S.C. 701 et seq.); the Beef Research and Information Act (7 U.S.C. 2901 et seq.); the Wheat and Wheat Foods Research and Nutrition Education Act (7 U.S.C. 3401 et seq.); the Animal Cancer Research Act (7 U.S.C. 3901 et seq.); the Floral Research and Consumer Information Act (7 U.S.C. 4301 et seq.); and the Forest Research Assistance Act (16 U.S.C. 582a-582a-7).

4

Major Biotech Institutions and Research Centres Around the World

Advisory Committee on Releases to the Environment

Home page for the UK Advisory Committee on Releases to the Environment (ACRE), which is part of the governments Department of the Environment, Transport and the Regions (DETR). ACRE gives advice to the government on issues concerning human and environmental safety concerning the release of genetically modified organisms (GMOs) and non-native organisms into the environment. It is provided by the Biotechnology Unit of the DETR, which is responsible for the operation and policy of the legislation controlling GMO release. This site provides the Committees agendas and reports, full-text newsletters, summary records for experimental releases and for marketing GMOs in Europe, information on international agreements, press releases and advice to ministers. Details of the Advisory Committee on Genetic Modification can also be found here. Published on the Web by the DETR.

Agricultural Biotechnology

The United States Department of Agriculture (USDA) is one of three Federal Agencies, along with the Environmental Protection Agency (EPA) and the U.S. Food and Drug Administration (FDA), primarily responsible for regulating biotechnology in the United States. Products are regulated according to their intended use, with some products being regulated under more than one agency.

AGROGENE

AGROGENE is a leader in plant biotechnology. The company's biotechnological expertise includes plant and animal breeding; fingerprinting and protection; genetic resources; industrial uses of plants; genetics; traceability and quality control; and GMO detection and quantification.

APS*net*-Plant Pathology On-Line

The international scientific organization American Phytopathological Society has been devoted to the study of plant diseases and their control since its founding in 1908. The society's members promote knowledge of plant diseases and their control through publications, meetings, symposia, workshops and special activities, as well as this web site.

Aurora Biosciences Corporation

Aurora Biosciences Corporation designs, develops and commercializes proprietary drug discovery systems and technologies to accelerate and enhance discovery of new medicines by the pharmaceutical and biopharmaceutical industries. The Company is developing a proprietary biology and high technology platform comprised of a portfolio of proprietary functional genomics technologies, and a highly automated ultra-high throughput screening system, known as the Company's UHTSS Platform. *http://www.aurorabio.com/*

Australian Biotechnology Association

The Australian Biotechnology Association is a national body of companies and individuals dedicated to the development and prosperity of the Australian biotechnology industry. It is the peak body for the Australian Biotechnology industry, and provides a 'platform' which brings together all the relevant players involved in the Australian biosciences community. Its mission is to facilitate the commercialisation of Australian bioscience in the international marketplace. This website provides information about the ABA, coming biotech events, and on-line access to a directory of Australian biotech companies and organisations, the ABA's journal "Australasian Biotechnology" and the ABA series of educational leaflets.

Babraham Institute

The Babraham Institute was founded in 1948 and is involved in biomedical research. This UK organisation is sponsored by the Biotechnology and Biological Sciences Research Council (BBSRC), and receives additional funding from medical charities and industrial organisations. The Institute's key areas of research are developmental genetics, immunology, neurobiology and cell signalling. The site provides information on the aims of the four major science programmes, the ongoing biological research taking place in the laboratories and the individual project leaders and their research. Additional features include a calendar of Babraham seminars, an electronic version of Babraham science essays, and access to the MEROPS database which provides information on proteases. For the general public the site also offers a "Science4All" section which covers topics such as the use of animals in research and the subsequent ethical considerations, the research areas covered by the Institute, an essay on developmental biology, and an archive of media press releases.

BBSRC

Ethics, Morality and Animal Biotechnology

Published on the Web by the Biotechnology and Biological Sciences Research Council (BBSRC), this paper was authored by Dr. Robert Straughan in September 2000. This document "examines the ethical, moral and social issues surrounding relatively recent developments that involve the genetic modification of animals, and cloning by nuclear transfer." Sections include: Current developments in animal biotechnology; Moral and ethical concerns; Animal ethics; Intrinsic concerns aabout animal technology; Entrinsic concerns aabout animal technology; and Conclusions. A technical annex is provided as are bibliographic references.

BINAS Online

Home page for the Biosafety Information Network and Advisory Service (BINAS), which is a service of the United Nations Industrial Development Organization (UNIDO). BINAS is responsible for monitoring global developments in regulatory issues in biotechnology. This site provides the BINASNews newsletter in full-text, which covers articles on topics such as genetically modified food and crops, and other socioeconomic issues within biotechnology. A calendar is also provided of events, information on regulations, discussion groups, and a library of publications.

Bio Online

Bio Online is a comprehensive site for biotechnology-related information and services. The site contains current news, featured articles, events, industry reports and analyses, press releases, product information, and much more. Bio Online exists to create a distributed on-line community of scientists, professionals, businesses and organizations supporting the life sciences for the purpose of facilitating communications and disseminating information within that community. Opening channels of information can only benefit those who strive to contribute to these endeavors.

BioABACUS

Biotechnology Abreviation and Acronym Uncovering

Service (BioABACUS) is a searchable database of abbreviations and acronyms in Biotechnology that contains terms in such categories as: Biochemistry, Cell Biology, Computers and Internet, Diseases, Grants, Journals, Laboratories, Medicine, Molecular Biology/Genetics, Neuroscience, Other Organizations, Professional Societies and US Government. The data provided include the name of the term, its meaning, the category in which it is mostly used and, when available, a link in which additional information for the term can be found.

BIOCODE

BIOCODE S.A. is a biotechnology company established in Liège, a city near Brussels, the capital of the European Economic Community. It specializes in the research, development, production and distribution of in vitro diagnostic kits (IVD). The company produces poly- and monoclonal antibodies in house and purifies proteins and peptides, runs immuno- and radioligand receptor assays as well as enzyme immuno- and DNA probe assays. BIOCODE's policy is to systematically upgrade and diversify its technologies and products in parallel with the progress of biotechnology and medicine.

BIOCOM/San Diego

BIOCOM/san diego is the regional association for biotechnology, medical device and bio-agriculture companies in San Diego County. It includes a significant service company sector, including finance, legal, architecture, marketing, insurance and other companies specializing in business with the life science industry. The company is actively engaged in ensuring that the life science industry remains a strong and growing part of the region's economy. BIOCOM proactively addresses significant business and legislative issues, educates the general public, and develops programs to help these companies operate efficiently and economically.

BioIndustry Association

Promoting UK Biotechnology

The BioIndustry Association (BIA) was founded in 1989 and currently represents over 200 small and medium sized biotechnology companies across the UK. The BIA represents the interest of this sector to government, the public and the European Union to "create a healthy UK bioscience sector which benefits society". The site gives a list of member companies, membership details, a list of events and publications, public affairs and news, details of the BIA's board of directors and their expert advisory committees.

BioInformatics Services (BIS)

BIS is a biotechnology firm specializing in applying the power of computational biology to build solutions to complex biological and data analysis problems faced by scientists in public and private research institutes, biotechnology companies, and universities.

Biology and Biotechnology Research Program at the Lawrence Livermore National Laboratory

The Biology and Biotechnology Research Program is focused on three strategic thrust areas: genomics, disease susceptibility and prevention, bioengineering research. These thrust areas provide a set of core compentencies and a foundation for our laboratory-wide strategic plan in the biosciences.

BioMetics, Inc.

BioMetics, Inc. provides engineering, design, validation, cGMP consulting and project management services for biopharmaceutical and environmental technology companies. The company delivers the highest quality of services, adding value to each project through innovative approaches to design and cost containment.

BioSpace.com

A global hubsite for the life sciences, including current news, industry events, resources, search tools, and much more, to help foster a worldwide community of shared interests within the biotech industry.

BIOTECanada

BIOTECanada is a national organization dedicated to promoting a better understanding of biotechnology and the many ways it contributes to improving the quality of life of all Canadians. BIOTECanada represents Canadian health care, agricultural, food, research and other organizations that are involved in biotechnology.

Biotech Medical Management Association (BMMA)

BMMA, a non-profit professional organization, establishes a forum for the open exchange of ideas and information between the pharmaceutical manufacturers and the health plan/payor community. The views of clinical provider, employer, consumer, and government representative members provide a balanced view on utilization management and therapeutic care standards. The goal of BMMA is to encourage active membership to impact and guide the future development of medical policy regarding new and evolving pharmaceutical products. With over 50% of the national population impacted by managed care and managed drug benefits, payors and providers are more actively involved in the approval and reimbursement of current and newly developed products.

BioTech's Biotechnology Dictionary

A searchable online dictionary of biology, chemistry and biotechnology terms. This free resource is still very much under construction. Currently, most of the 8300+ terms deal with biochemistry, biotechnology, botany, cell biology and genetics. There are also some terms relating to ecology, limnology, pharmacology, toxicology and medicine. Don't expect to find common or exotic animals here; there are far too many species to cover even a fraction in a resource like this. However, medically- and biotechnologically-relevant organisms such as bacteria, worms, fungi, and some plants are included. Looking for a chemical acronym? Try searching BioTech's Chemical Acronyms Database.

Bio-Technical Resources

Bio-Technical Resources provides innovative solutions for the development of biotechnology products and processes through contract services in research and development, bioprocess scale-up, pilot scale manufacturing, technology and economic assessments.

Biotechnology and Applied Biochemistry

Biotechnology and Applied Biochemistry publishes research concerning the expression, purification, characterisation and application of biological macromolecules in therapeutics and diagnostics. Tables of contents and abstracts are available from 1987, with full-text articles available to subscribers from 1998. Instructions to authors and subscription details are provided. Published on the Web by Portland Press on behalf the of International Union of Biochemistry and Molecular Biology (IUBMB).

Biotechnology and Biological Sciences Research Council (BBSRC)

The BBSRC is the leading funding agency for academic research and training in the biosciences at universities and institutes throughout the UK. The research that funded aims to improve the understanding of living organisms and biological systems in seven major areas: agri-food, animal sciences, biochemistry and cell biology, biomolecular sciences, engineering and biological systems, genes and developmental biology, and plant and microbial sciences.

Biotechnology Group of Danish Institute of Agricultural Sciences

Biotechnology Group of Danish Institute of Agricultural Sciences, DIAS, is a small research team engaged in plant breeding related research. Promoting the use of transgenic crop plants in agriculture is at the core of all research activities of the group. All projects comprise a mix of applied and fundamental science, and the Group is thus an important link between applied oriented DIAS and the Danish universities.

Biotechnology Industry Organization

This international trade association, also known as BIO, represents companies, academic institutions, and state biotechnology centers. Its members are involved in the research and development of health care, agricultural, industrial, and environmental biotechnology products. BIO's web resources include a guide to biotechnology for editors and reporters and information about biotech issues and policies regarding food and agriculture, bioethics, and biotech drugs.

Biotechnology Information Resource

The Biotechnology Information Center (BIC) is one of ten information centres at the National Agricultural Library located in Beltsville, Maryland. The Biotechnology Information Center provides access to a variety of information services and publications covering many aspects of agricultural and environmental biotechnology, including transgenic animals, current gene mapping projects, legislation and regulation, patenting issues and biotechnology newsletters. There are also some collections of resources on more specific topics, such as bovine somatotropin (bST/BGH), bioremediation and Bacillus thuringiensis. It should be noted that the Biotechnology Resource Web site is no longer updated and is useful for general purposes only.

Biotechnology Program at UC Davis

UC Davis has long enjoyed a reputation as one of the world's leading institutions for research in biological sciences. UC Davis houses leading departments in each of the major areas where biotechnology has a significant impact: agriculture covering plant, animal, microbial science, entomology and sustainability; medicine and veterinary medicine including therapeutics, diagnostics and pathology; food science; bioprocess technology and environmental science. Research in molecular and cellular biology and biochemistry has led to the development of new research approaches and the new tools of biotechnology. UC Davis views these recent research results as part of a progressive cycle, with technology contributing to the advancement of research goals and biological research contributing, in turn, to the advancement of technology.

Biotechnology Tutorial

A tutorial that covers the the fundamentals of applied molecular biology and biotechnology. The tutorial includes illustrated lecture notes, exams and quizzes. It is aimed predominantly at graduate students in the United States, and is produced by Dr Michael Blaber at the Department of Chemistry, Florida State University.

Biotechnology

A series of documents about biotechnology aimed at anyone with an interest in the field. Sections covering the topic include: regulations and guidance on safety assessments, food labelling, FDA presentations and testimony about bioengineered foods, information for consumers, and compliance. Published on the Web by the US Food and

Drug Administration, Center for Food Safety & Applied Nutrition.

Biotechnology; CAMR

CAMR (Centre for Applied Microbiology & Research) is a centre for research and development and production of biopharmaceuticals, and the improvement of the environment through its expertise in microbiology. Their aim is to "contribute to the health of the UK population by conducting research on specified microbiological hazards with a view to the development and production of effective diagnostic, prophylactic and therapeutic products." As part of the Microbiological Research Authority, CAMR reports to the Secretary of State for Health. The site provides details of vacancies, news and events, GMP manufacturing, contract research, partnership opportunities, process development, as well as linking to the European Collection of Cell Cultures (ECACC) Web pages.

BioValley

The principle goal of BioValley is the promotion of greater cooperation between companies involved in the biotechnological and biomedical sectors and the internationally respected scientific institutions associated with universities in the BioValley area, most of which have already formed close links between themselves. The intention is to address not only the many pharmaceutical concerns already present in the BioValley area, but also to involve the region's numerous smaller enterprises and suppliers. This calls for the creation of a network strongly committed to technology transfer.

BIRD Engineering

Biotechnology Integrated Research, Development & Engineering (BIRD Engineering) specialises in contract research and engineering in the areas of the fermentation technology and downstream processing. BIRD Engineering was founded by the combined initiative of researchers working in the field of bioprocess engineering, industrial microbiology and biochemistry at Delft University of Technology and Applikon Dependable Instruments, manufacturer of fermentation equipment and analytical instruments. BIRD was originally set up to bridge the gap between research at universities and industrial activities in biotechnology. The company not only performs the necessary research but can also the development and engineering studies with respect to industrial biotechnological applications. In this way, BIRD comes up with solutions for your problem that is not just an academic one but has also a practical value. The main focuses are on those application areas which require a combined know-how of bioprocess technology, microbiology, biochemistry and genetics. In this central position BIRD is able to assist from the selection of micro-organisms to scale-up of production processes.

Boston Biomedica, Inc.

Boston Biomedica, Inc. is an infectious diseases management company providing products and services for the detection and treatment of infectious diseases (ID) such as AIDS, Lyme Disease, and Viral Hepatitis. Boston Biomedica is also pursuing research and development programs in other areas, including Pressure Cycling Technology (PCT), a new technology which may allow more precise control chemical of reactions through the use of pressure.

Boston Life Sciences, Inc.

Boston Life Sciences, Inc., is a development stage biotechnology company engaged in the research and development of novel therapeutic and diagnostic products to treat chronic debilitating diseases such as cancer, central nervous system disorders and autoimmune diseases. The Company currently has 11 technologies in its product portfolio, the majority of which were invented or discovered by researchers working at Harvard and its affiliates and have been licensed to the Company.

Calendar of Genome and Biotechnology Meetings

This calendar is intended to inform genome researchers about pertinent genome-related meetings. Access is also given to a list of training events and workshops of interest to genome researchers. Listed under the coming months are a list of meetings. Links are also given to Biocomputing Courses available. This Web site is maintained by the Human Genome Management Information System (HGMIS) at Oak Ridge National Laboratory for the US Department of Energy Human Genome Program, and is sponsored by the US Department of Energy Office of Science, the Office of Biological and Environmental Research, and the Human Genome Program.

California Healthcare Institute

The mission of California Healthcare Institute is to (1) research, develop, and advocate policies and actions that promote biomedical science, biotechnology, pharmaceutical and medical device innovation in California; (2) create a favorable environment for the State's biomedical and health care technology community to discover, produce, and deliver products that benefit society; (3)provide a forum to identify, analyze, and develop positions on public policy issues that affect California's biomedical and health care technology interest; (4) conduct research as a basis for advocating responsible state and federal policies; (5) communicate to public officials, providers, patients, and the general public the value of the State's health care technology community for the health and economic well-being of our nation's citizens and the economic growth of California; and (6) identify and communicate the value of the products of health care technology with regard to cost, benefit, and patients' quality of life.

Cambridge Healthtech Institute (CHI)

CHI services the biotechnical and biomedical markets by acting as an information resource in an ever-changing environment. Through the organization of scientific conferences and publication of market reports, CHI offers a one-stop resource for learning about the latest technological developments in such areas as genomics, drug discovery and development, lab instrumentation, and blood products. CHI holds over 50 U.S. and international events each year, and currently offer over 12 market reports in its new Genomic Pathways Report Series. CHI events facilitate interaction with key decision makers from leading pharmaceutical and biotech organizations. This format offers companies who sponsor or exhibit at a CHI event direct access to individuals who are interested in their products and services.

Capralogics Inc.

Capralogics provides custom antibody services to the research community and is a manufacturer of polyclonal antibodies against cell membrane receptors, nuclear components, apoptosis factors, neuropeptides, oncoproteins, signaling intermediates, transcription factors and proteins involved in infectious diseases. The company offers a complete service package ranging from protocol and peptide development through purification services.

Cascade Biologics Inc.

Cascade Biologics is a supplier of cell culture and molecular biology products. The scientists at Cascade Biologics are experts in the isolation and propagation of numerous cell types from different tissues, and in the development of specialized cell culture media for the growth of these cells.

CEDRA Corporation

CEDRA Corporation is a provider of excellence in biotechnical services for the pharmaceutical industry and federal government. In addition to analytical chemistry services, CEDRA also offers toxicology/pharmacology, metabolic, synthetic chemistry, and pharmacokinetic/statistical services.

Cell Trends Inc.

Cell Trends Inc. is a biotechnology firm that offers technical and research project work to scientists of virtually every level of sophistication: industry, government agencies, medical schools, colleges, and universities. Cell Trends offers a wide range of cell based laboratory services, utilizing a diverse selection of mammalian and insect cell lines. Features include baculovirus gene expression, DNA recombinant technology, soluble and membrane bound protein production (using hollow fiber bioreactors, roller bottles, or spinner flasks), gene cloning, transformant selection, electrophoresis, PCR, ELISA, and selection of immortalized cell lines.

Center for Advanced Biotechnology and Medicine (CABM)

CABM was established in response to a recommendation in the report of the Governor's Commission on Science and Technology that New Jersey should create and maintain a world-class center in biotechnology. Launched in 1985, CABM now encompasses the Core Facility, conducting basic research in the areas of structural biology, molecular genetics, and cell and developmental biology; and educational Programs, including lectures, seminars, symposia, short courses, fellowships, and visiting scientist programs.

Center for Advanced Research in Biotechnology (CARB)

CARB, located at the University of Maryland, was designed to meet the specific needs of the biotechnology community in the field of protein structure, function, and design. Scientists from the Biotechnology Division of the Chemical Science and Technology Laboratory of NIST and the University of Maryland have come together to conduct research programs in structural and physical biochemistry and molecular biology. CARB, with sophisticated state-of-the-art instrumentation, provides an exceptional environment for training graduate students and postdoctoral fellows in fields critical to the future of biotechnology. CARB offers a Visiting Scientists Program that allows researchers from industrial, academic, and governmental laboratories to work at the facility for extended periods.

Center for Biotechnology at Oak Ridge National Laboratory

Biotechnology is an integrating science that emphasizes the development and application of biological approaches, or the use of engineering and computational capabilities, to meet national needs in health and environmental protection, biomedical applications, environmental remediation, pollution abatement, and energy production. ORNL is uniquely positioned in this emerging field owing to its diversity of interactive disciplinary foundations (e.g., biology, biochemistry, molecular biology, computational sciences, chemistry, ecology, bioengineering, and biophysics) and its core competency in biological and environmental science and technology. The ORNL Center for Biotechnology provides a coordinated focus for this biotechnology program, integrating activities across 14 ORNL divisions and forming partnerships with other institutions, both public and private, throughout the world.

ChemGen

ChemGen is a privately held biotechnology company specializing in developing products through the use of microbial fermentation technology. The company's focus has been on the production of enzymes for patented applications and on chemicals, including pharmaceutical intermediates, using unique production processes.

ChemGenes Corporation

ChemGenes Corporation, a biotechnology company engaged in the field of DNA/RNA synthesis. ChemGenes holds a number of patents in the field of standard, modified, as well as backbone modified DNA/RNA synthesis products.

Cleveland Biotech Ltd.

Cleveland Biotech grows and acclimatises naturally occurring bacteria to create products for the degradation of organic wastes, including fats, oil and greases, cellulose, petroleum hydrocarbons, and chlorinated organics. Products are also available for the degradation of inorganic pollutants such as ammonia and nitrite.

Collateral Therapeutics

Collateral Therapeutics is focused on the discovery, development and commercialization of non-surgical gene therapy products for the treatment of cardiovascular diseases, including coronary artery disease, peripheral vascular disease, congestive heart failure and heart attack. The Company's non-surgical gene therapy product candidates include GENERX, GENEVX, GENVASCOR, GENECOR, CORGENIC, and MYOCOR. These non-surgical gene therapy products use technology, which includes methods of gene therapy, therapeutic genes, and gene delivery vectors.

Commonwealth Biotechnologies, Inc.

Commonwealth Biotechnologies provide sophisticated

research and development support services on a contract basis to the biotechnology industry. The Company's customers consist of private companies, academic institutions and government agencies, all of which use biological processes to develop products for healthcare, agricultural and other purposes. The Company's revenues are derived principally from providing macromolecular synthetic and analytical services to researchers in the biotechnology industry or to researchers who are engaged in life sciences research in government or academic labs throughout the world.

Compugen

Compugen is a pioneer in the fields of computational genomics and proteomics. The company combines the disciplines of mathematics and computer science with molecular biology to improve the understanding of genomics and proteomics, the study of genes and proteins. Compugen's expertise in these areas is used to develop core technologies and platforms that are the basis for its products and services. These products and services are designed to enable life scientists to significantly enhance their research efforts in drug discovery, therapeutic and diagnostic. Compugen provides innovative products and services to pharmaceutical, biotechnology and other life science organizations through corporate solutions and to individual life scientists. In addition, the company is commercializing the genes and proteins that are discovered through the company's drug discovery division.

Contained Use of Genetically Modified Organisms

This leaflet summarises the controls on the contained use of genetically modified organisms, set out in the Genetically Modified Organisms (Contained Use) Regulations 2000. These Regulations are supplemented by the sections of the Environmental Protection Act 1990 and the Genetically

Modified Organisms (Risk Assessment) (Records and Exemptions) Regulations (as amended in 1997) which specifically cover the control of risks to the environment from genetically modified animals and plants. The leaflet lists what the regulations cover, what is meant by contained use, what the main duties are under the regulations, who enforces the regulations, how the work is classified, notifications and consents, how notification should be made, and fees. It also outlines in brief the responsibilities of other Government organisations dealing with biotechnology, giving their addresses and telephone numbers for further information. This leaflet is published by the Health and Safety Executive and was added to the Web in November 2000.

Council for Biotechnology Information

Leading biotechnology companies have founded Council for Biotechnology Information to create a public dialogue and to share information about biotechnology that is based on objective scientific research, independent expert opinion, and peer-reviewed published reports. The web site is also available in the French language.

Council for Responsible Genetics

The Council for Responsible Genetics (CRG), established in 1983, is a US non-profit organisation of scientists, public health advocates, and others. Their main aim is "to promote a comprehensive public interest agenda for biotechnology." Online information available on this site includes genetic discrimination, cloning, patenting of life forms, and genetically engineered food. GeneWatch is a bi-monthly bulletin dedicated to the social and environmental implications of biotechnology, and selected articles are available online from January 1999. Other resources which are found on this site are CRG alerts on biotechnology development and an online petition. This site is aimed at anyone who is interested in the impact of new biotechnologies and their sociological and environmental impact.

Covance Inc.

Covance's biotechnology services include capabilities in cGMP manufacture of monoclonal antibodies, growth factors, cytokines, and other recombinant products for clinical trials and commercial use. The company offers a comprehensive range of experience developing biopharmaceutical production processes leading to optimal expression, purification, formulation, and testing of recombinant products. Equally important, Covance can effectively transfer established manufacturing processes into large-scale cGMP manufacturing areas and provide a reliable supply of the highest quality recombinant product.

Cranfield Biotechnology Centre, Cranfield University

Established in 1981, Cranfield Biotechnology Centre is an integral part of the Institute of BioScience and Technology at Cranfield University. Information is provided on the Centre, its staff, research interests and projects focusing in three areas; medical diagnostics, environmental diagnostics, and food biotechnology. Details are given of postgraduate training in biotechnology, biomedicine and analytical science, and current vacancies within the Institute are advertised. Information on the Institute's commercial activities are also available.

CSL Limited

CSL Limited is an Australian public company specialising in the development, manufacture and marketing of biologically based health care products which benefit the community. Products include human and veterinary pharmaceuticals (notably vaccines) and products derived from human plasma, diagnostics, and cell culture reagents.

CSL is firmly committed to research and development, quality assurance and the development of international product and marketing alliances.

CuraGen Corporation

CuraGen Corporation is an integrated, genomics-based drug discovery and development company that is revolutionizing the Life Sciences industry by merging biology with information technology. CuraGen has created an integrated suite of genomics technologies to better understand the role of genes in improving human health, animal health and agricultural vitality. This technology is being applied to augment the discovery and development efforts of the Company's collaborators, while simultaneously advancing the development of CuraGen's internal pipeline of protein therapeutics and antibody drugs for the treatment of complex diseases.

Current Opinion in Biotechnology

The electronic version of the journal Current Opinion in Biotechnology, published by Elsevier Science Ltd. The journal covers plant biotechnology, biochemical engineering, pharmaceutical and environmental biotechnology, food biotechnology, analytical biotechnology, and expression vectors and delivery systems. The site contains information for authors, details on electronic submission of manuscripts, the make-up of the editorial board, and subscription rates. Tables of contents and abstracts are available free of charge to any registered BioMedNet user, with full-text of articles available in PDF (requires Adobe Acrobat Reader) to licensed institutions or users with personal subscriptions. The journal is published online by Elsevier Science Ltd.

Department of Bioscience, University of Strathclyde

The home page for the Department of Bioscience at the University of Strathclyde. The site provides general information about the department, as well as undergraduate courses offered, postgraduate opportunities, positions available, research facilities, and a biotechnology section for schools. Details of the Department's principal investigators, their research interests and contact details are available. The research interests include glycobiology, molecular microbiology, molecular signalling, biomedical microbiology, microenzymes, and microbial physiology and fermentation technology.

Department of Biosciences, University of Kent

Home page for the Department of Biosciences and the associated Research School of Biosciences (RSB) at the University of Kent at Canterbury. The site provides general information about the department, undergraduate courses offered and postgraduate opportunities. Details of the RSB's principal investigators, their research interests and contact details are available. The research interests of the School fall into four principal areas: Protein Science, Microbiology, Cell Biology, and Clinical Biosciences.

Division of Environmental and Applied Biology, University of Dundee

Home page of the Division of Environmental and Applied Biology within the School of Life Sciences at the University of Dundee. The overall research of the Division is directed towards how living organisms respond and adapt to changes in the environment caused by both natural and industrial effects. Information is provided on the staff and their current research interests, including basic research in microbial, plant and animal ecophysiology and molecular biology, and environmental biotechnology. Details are provided of postgraduate training within the School of Life Sciences, including PhD programmes and MRes courses, and a list of current vacancies within the School is available. A News section covers recent events, both scientific and general, of interest to the School.

Dyadic International Inc.

Dyadic is a leader in the development and production of industrial enzymes and speciality chemicals, holding US and International patents. Currently the company's *non-GMO* products are used in the textile, food, starch and animal feed industries. An area targeted for future development is the pulp and paper industry.

Dynal Biotech

Dynal Biotech, a company in Oslo, Norway, developed Dynabeads, which are magnetic beads for efficient and quick separation of cells, bacteria, proteins, subcellular compartments, mRNA and DNA. This development has revolutionised the isolation and separation of many biological materials. For instance, the attachment of target-specific antibodies to the surface of the beads allows capture and isolation of intact cells directly from a complex suspension, such as blood. This is all accomplished under the influence of a simple magnetic field without the need of column techniques or centrifugation. Dynabeads have been applied to three major scientific areas: immunology, molecular biology, and microbiology. Dynal Biotech is now taking involvement in cancer research areas, such as the early detection of minimal residual cancer. In addition, the company continues to enhance its competence in its core technology; particle structure and surface chemistry, affinity ligands and immobilisation strategies combined with the on-going development of dedicated products as a solution provider in the biological and biomedical sciences.

EJB

Electronic Journal of Biotechnology: The home page of the Electronic Journal of Biotechnology (EJB), a scientific peer-reviewed journal edited exclusively in electronic format and published free on the Web. EJB publishes review and research articles in full-text, available in either HTML or PDF (requiring Adobe Acrobat Reader), with an archive dated back to volume 1, 1998. Other areas covered by the journal include molecular biology and genetics, policies on international cooperation, and plant microbiology. Articles can be searched for by either author, keyword and area. EJB is supported by the UNESCO / MIRCEN network. Published on the Web by the Universidad Católica de Valparaíso, Chile.

Enchira Biotechnology Corporation

Enchira Biotechnology is a leading-edge biotechnology company incorporating genetic recombination and high throughput screening in an integrated, directed evolution technology platform. This proprietary platform technology can be used to generate libraries of novel genes for the creation of improved gene-based products for a broad range of applications, such as protein-based pharmaceuticals, agricultural crop enhancement and protection products, and industrial enzymes for manufacture of specialty chemicals, fine chemicals and pharmaceutical intermediates. The ability to generate novel proteins with enhanced or altered properties has stimulated interest in all areas of biotechnology, medicine, and the chemical sciences.

Enviro Research Pty Ltd

Enviro Research has been engaged in microalgal culture research for a number of years. The company has developed the world's first and only closed photobioreactor which can be scaled up to any volume. The Envirocell offers all the control, hygiene, efficiency and reliability of laboratory based culture systems at industrial volumes. The Envirocell is currently the best microalgal culture system in the world for the aquaculture industry, fine chemicals/pharmaceuticals companies, health food/nutraceuticals industry, and wastewater nutrients/toxins removal industry.

Epimmune Inc.

Epimmune develops therapeutic products for treatment of immunological diseases. The company's core business is the development of vaccines to treat and prevent infectious diseases and cancer. Epimmune's business also includes a therapeutic anti-inflammatory program based on cell adhesion inhibitors and its Glytec business unit, encompassing proprietary manufacturing technology for the enzymatic synthesis of bioactive complex carbohydrates.

Europa

Biosociety

"Biosociety" aims to stimulate "interactions between biotech researchers and social scientists, economists and ethical experts active in the field of life sciences." Topics discussed include life sciences for health, food quality and safety, bioethics, and public understanding. There is also information on policy aspects, research projects, a document library, an "expert" database, and a news and events section. A Bioforum offers "an arena for open and pluralistic debate on critical Biosociety-related issues." Biosociety is part of Europa, the portal site of the European Union that provides up-to-date coverage of European Union affairs and essential information on European integration. Made available on the Web by the European Commission.

European Federation of Biotechnology

The European Federation of Biotechnology is a voluntary association of European non-profit-making scientific and technical societies active in the field of biotechnology. The aim of the association is to enhance cooperation between these societies in order to promote R&D in the multidisciplinary field of biotechnology and to support its scientific, economic and applied technical development in a socially and ethically acceptable manner. Also see their Agri-Biotechnology Section.

European Initiative for Biotechnology Education

The European Initiative for Biotechnology Education (EIBE) seeks to promote skills, enhance understanding and facilitate public debate throughout Europe. The group provides teaching materials, protocols and practical activities suitable for both schools and research. EIBE has become an active European multidisciplinary network of experts in biotechnology education drawn from 20 centres in 17 European countries. Users require Adobe Acrobat Reader to read and print PDF files. The site is also available in French, German, Dutch and Italian, as well as English.

Exiqon A/S

Exiqon is a Danish biotech company focused on the development of enabling technology platforms and novel DNA- and immuno-diagnostic products against common genetic diseases, in particular, cardio-vascular, diabetic and other chronic and degenerative diseases. In other fields Exiqon is a potential technology partner for licensees. Exiqon's IPR includes LNA (Locked Nucleic Acids) and a universal method for the covalent coupling of diagnostic molecules to solid supports.

Federal Bio-Technology Transfer Directory

The Federal Bio-Technology Transfer Directory is an integrated information resource designed for users in the biotechnology, pharmaceutical and healthcare industries. Besides being essential for access to technology transfer opportunities, the database is a unique source for competitive intelligence concerning federal (and corporate) research, development and commercialization activities.

Gemini Biotech

Gemini Biotech provides prompt, high quality, and cost-effective support for research and drug discovery. The company's technical staff and consultants are experts in synthetic chemistry and molecular biology. Gemini Biotech specializes in custom DNA probes/primers, modified oligonucleotides and DNA-RNA hybrids, full-length genes, gene sequencing, custom peptides and modifications, peptide conjugates, and peptide antibodies.

Genetically Modified Crops

The Ethical and Social Issues: A report from a Working Party at the Nuffield Council on Bioethics, on the ethical and social issues surrounding GM technology. Chapters are included on the scientific and ethical principles of genetic technologies, the scientific basis of genetic modification, issues related to commercial implementation, the impact on developing countries, consumer choice and food quality, the environmental impact, policy, and conclusions and recommendations. Published in May 1999, the report is provided in HTML format and in PDF (requiring Adobe Acrobat Reader). Made available on the Web by the Nuffield Foundation, an independent charity providing grants for research and innovation.

Genetics Forum

The Genetics Forum is an independent organisation in the UK concerned with the use of new genetic technologies and their public policy implications. It was founded by a group of scientists, lawyers and advocates from the animal welfare, environmental and consumer movements concerned about the long-term impact of rapid developments in the genetic sciences.

Genetics Institute

Genetics Institute is a leading biopharmaceutical company engaged in the discovery, development and commercialization of human pharmaceuticals through recombinant DNA and other technologies. The company has a diversified portfolio of licensed and proprietary pharmaceutical products at various stages of development, including treatments for anemia, hemophilia, cancer, bone damage, inflammatory conditions and immune system disorders.

GeneWatch UK

GeneWatch UK is a non-profit policy research and public interest group. The areas of interest are divided into the following sections: genetically modified (GM) crops and food, human genetics, GM animals, laboratory use of GMOs, patenting, and bio-weapons. A comprehensive archive of publications relating to these topics is provided. A database of information on GM crops and food is also available, which was developed in assistance with a grant from the Greenpeace Environmental Trust. GeneWatch is aimed at anyone who has an interest in the ethical issues raised by genetic engineering. This Web site has been included in the Biotechnology section of the Nature Net Guide as an example of "what impressive things can be done in communicating and facilitating research".

Genomes to Life

The US Department of Energy's (DOE) Genomes to Life (GTL) programme aims to use the new genomic data and high-throughput technologies for studying the proteins encoded by the human genome to explore the diverse natural capabilities in microbes. This site provides the DOE's mission challenges, underlying goals, current research, publications and images, enabling technologies, examples of science enabled by GTL, and future benefits. Genomes to Life is made available on the Web by the Human Genome Management Information System (HGMIS), and is sponsored by the Office of Biological and Environmental Research and Office of Advanced Scientific Computing Research of the US Department of Energy Office of Science.

Genzyme Transgenics Corporation

Genzyme Transgenics Corporation, a biotechnology company, is a leader in the production of therapeutic proteins in the milk of transgenic animals. Transgenic animals carry specific genetic material that allows them to produce human proteins in their milk. Unique molecular biology and production expertise has enabled Genzyme Transgenics to produce more than 60 therapeutic proteins, including monoclonal antibodies, plasma proteins and certain hard-to-express proteins in the milk of mice, rabbits, goats and cows.

The company has developed substantial expertise and increasing focus on monoclonal antibodies for therapeutic use in chronic diseases such as rheumatoid arthritis, other autoimmune conditions and cancer. *http://www.transgenics.com/*

Graphics Gallery

The Graphics Gallery is a series of annotated illustrations representing important processes relevant to biotechnology. Each diagram is provided with a summary of information. The diagrams are divided into the following groups: from gene to function, building blocks, chromosomes and cell division, biological engineering, genetics, viruses, and cell processes. The Graphics Gallery is part of the "About Biotech" section of Access Excellence, a US national educational programme "that provides high school biology and life science teachers access to their colleagues, scientists, and critical sources of new scientific information via the Web." Made available on the Web by the US National Health Museum.

Greenpeace

Genetic Engineering: This site, which is operated by Greenpeace, advocates no genetic manipulation of nature. It focuses on (1) promoting natural foods and the segregation of genetically engineered foods with proper labels, (2) preventing genetically modified organisms from being released into the environment, (3) preventing any living organisms or their genes being subject to private property claims (i.e., no patents), and (4) promoting genetic engineering issues being addressed by international bodies, with international regulations based on the precautionary principle. Greenpeace USA, a part of the international organization dedicated to protecting the health of planet Earth, offers on-line media kits providing background information on various issues, including Greenpeace's concern that the FDA's safety standards regarding genetically engineered foods are insufficient and may be putting human health at risk. This media kit links to Greenpeace's report titled *FDA Approved: Genetically Engineered Soy and Corn in Baby Food and Nutritional Supplements* and to additional related information. *http://www.greenpeace.org/~geneng/gehome.htm*

Igene Biotechnology

Igene Biotechnology focuses on the commercialization of value-added specialty biochemical products for the food, feed, flavor and agrichemical industries. Through the use of state-of-the-art fermentation technology, physical and chemical separation technology, and related chemical and biochemical engineering technologies, the company has developed proprietary processes to convert selected agricultural raw materials or feedstock into commercially useful and cost effective "environmentally friendly" natural alternatives to products now produced by chemical synthesis. *http://www.igene.com/*

InfoBiotech Canada (IBC)

IBC is a partnership of government, private and academic sectors with the goal of providing enhanced access to information on Biotechnology in Canada and worldwide.

Information Systems for Biotechnology (ISB)

ISB provides information resources to support the environmentally responsible use of agricultural biotechnology products. Here you will find documents and searchable databases pertaining to the development, testing and regulatory review of genetically modified plants, animals and microorganisms within the U.S. and abroad.

Insect Biotech Canada Inc. (IBC)

Insect Biotech Canada Inc. (IBC) is a Canadian-based, non-profit biotechnology research and development consortium focused on the application of biotechnology to insects and their pathogens. IBC has brought together Canada's top researchers in insect physiology, biochemistry, insect virology and molecular biology in an integrated and coordinated organization. The network is focused on serving the needs of Canada and the international community in identifying and controlling insect pests using current techniques of biotechnology.

Institute of Biosciences and Bioengineering, Rice University

The Institute's mission is to promote cross-disciplinary research and education encompassing the biological, chemical, and engineering disciplines. The Institute represents a unique educational environment in which to prepare students to meet the complex scientific, technological, and ethical challenges they face as they head into the twenty-first century.

Institute of Biotechnology, University of Cambridge

Home page of the Institute of Biotechnology at the University of Cambridge, giving access to information on staff and their current research activities, research facilities, PhD and MPhil opportunities, and positions available within the Institute. Established in 1988, the Institute has a multi-disiplinary approach and currently has expertise in the following areas: molecular biology, metabolic engineering, protein engineering, microbiology, biotransformations, biodegradation, plant biotechnology, enzyme technology, bio-separations, chemical sensors, biosensors, instrumentation, combinatorial chemistry, and molecular modelling.

International Centre for Genetic Engineering and Biotechnology

ICGEB: The International Centre for Genetic Engineering and Biotechnology (ICGEB) is a statutory body founded in 1983 through an initiative led by the United Nations Industrial Development Organization (UNIDO). In the beginning ICGEB operated as a special programme of UNIDO, gaining status in 1995 as an independent, fully autonomous international, intergovernmental organisation.

The major goal of the ICGEB is to advance research and training in molecular biology and biotechnology, particularly in the developing world. ICGEB operates from two main centres, Trieste, Italy and New Delhi, India.

This Web site provides details on affiliated centres (46 countries are currently full members), continuing research and ongoing activities, fellowships, grants and sponsored training courses, and web resources for biosafety, biotechnology transfer, patents and a computer resource for molecular biology called ICGEBnet.

International Centre for Genetic Engineering and Biotechnology

The International Centre for Genetic Engineering and Biotechnology (ICGEB) was established in 1987 by the United Nations Industrial Development Organization (UNIDO) to promote the safe use of biotechnology worldwide and with special regard to the needs of the developing world. In 1995 it became a fully autonomous international organization composed of 41 member states. It has two main laboratories: one in Trieste, Italy, and one in New Delhi, India. It also hosts a network of 27 Affiliated Centres, which are national laboratories of member states. It sponsors both basic and applied research, including the production of novel malaria and hepatitis vaccines, the study of human pathogenic viruses, the study of human genetic diseases, and the genetic manipulation of plants. In 1995, ICGEB and UNIDO created BINAS (Biosafety Information Network and Advisory Service) to monitor global developments in regulatory issues in biotechnology. In 1998, the Centre added the ICGEB Biosafety WebPages, which are totally dedicated to biosafety and risk assessment for the environmental release of genetically modified organisms. Also in 1998, the Centre began publishing *biosafety*, a bibliography of scientific articles on the main topics of concern about the environmental release of genetically modified organisms.

InterSciences Inc.

InterSciences is a privately-owned and operated, scientific instrumentation and supplies marketing organization representing specialized principal manufacturers lines. InterScience's primary marketing focus is Biotechnology for academic and hospital research, industrial, government, private lab and Pharmaceutical segments of the Canadian market. *http://www.interscience.com/*

Iomai Corporation

Iomai is a biopharmaceutical company with products that greatly enhance human immune systems in the fight against disease using its Transcutaneous Immuno Modulation (TIM) technology for immunoprevention and immunotherapy. TIM works through the skin as a non-invasive booster to the body's own immune system and can be targeted against specific diseases. TIM products are administered easily and safely — painlessly stimulating the immune system to prevent or treat disease through a skin patch, without needles or injections. The Company is focused on rapid commercialization of the TIM technology through strategic relationships with industry partners. The TIM project pipeline includes products for three target markets: infectious diseases, cancer therapies and autoimmune therapies. *http://iomai.com/*

Journal of Molecular Microbiology and Biotechnology

The home page for the Journal of Molecular Microbiology and Biotechnology (JMMB). JMMB publishes papers in all aspects of molecular microbiology and molecular biotechnology. It features written symposia, reviews, correspondence, and information for authors. Free access is provided from August 1999 (vol.1, no.1) to tables of contents and abstracts, with full-text access available to subscribers only. A free sample issue is available. Made available on the Web by Horizon Scientific Press.

Laboratory Biosafety Manual

Since it was first published in 1983, the Laboratory Biosafety Manual has provided practical guidance on biosafety techniques for use in laboratories at all levels. Published in

2004, the new third edition of the manual has been extensively revised and expanded. The manual now covers risk assessment and safe use of recombinant DNA technology, laboratory biosecurity concepts are introduced, and the latest regulations for the transport of infectious substances are reflected. Areas covered include: Biosafety Guidelines; Biosecurity Guidelines; Laboratory equipment; Good microbiological techniques; Introduction to biotechnology; Chemical, fire and electrical safety; Safety organization and training; and a Safety checklist. Made available on the Web by the World Health Organization (WHO).

Leinco Technologies

Leinco Technologies is a global biosciences company with manufacturing facilities. The company is built on a diverse biotechnology platform and offers product development and contract manufacturing services under GLSP and cGMP for active biopharmaceutical intermediates. The company also offers an extensive line of antibodies, recombinant proteins and reagents for research in immunology, cell biology, and neuroscience. *http://www.leinco.com/*

Life Science Dictionary

This illustrated glossary concentrates mainly on terms associated with biochemistry, biotechnology, botany, cell biology and genetics, as well as some terms relating to ecology, limnology, pharmacology, toxicology and medicine. There are short definitions for each term, a search facility and an alphabetical listing of all the entries (over 8300). Made available on the Web by the BioTech Resources Web Project at the University of Texas Institute for Cellular and Molecular Biology.

Life Sciences Gateway

The Life Sciences Gateway is published on the Web by

the Life Sciences Branch of Industry Canada, which aims to "advance the economic growth and competitiveness of Canada's life sciences sector." This site provides access to a range of information products and services on all aspects of biotechnology, as well as emerging technologies such as bioinformatics, genomics & proteomics, and nanotechnology. The site also provides an overview of the Canadian biotechnology industry, outlines the capabilities of Canadian biotech companies, details Canadian government programmes and strategies, and has information relating to the international business of biotechnology.

LifeSciencesWorld

LifeSciencesWorld (formerly known as Biotechfind.com) is a directory of international links covering the fields of biotechnologies, including the latest biotechnology news and job vacancies. Information is provided on free biotechnology publications and upcoming events and conferences. A searchable Career Centre is available for use by both job seekers and employers. The LifeSciencesWorld Web site is also available in French. LifeSciencesWorld is managed by Synapse Information Technologies.

Link Technologies Ltd

Link Technologies Ltd, established in Glasgow in 1989, is a growing biotechnology company that now operates at the Strathclyde Business Park. The company, which has a strong reputation in the U.S. and European markets for manufacturing quality chemical reagents for the synthesis and modification of DNA and RNA, is now also developing innovative reagents for use in related areas of molecular biology, such as the structural study of proteins and the synthesis of oligonucleotide-peptide conjugates. In tandem with its expanding catalogue of products, the company offers a contract synthesis service for molecular biology reagents.

Louisiana Alliance for Biotechnology (LAB)

LAB is dedicated to enhancing biotechnology research and business development in the State of Louisiana. The goals set for LAB in order to accomplish this mission are to facilitate: the development of an interactive consortium of university and private sector researchers in molecular biology and biotechnology; the transfer of scientific advances into the private sector; enhancement of the university academic environment and community quality of life through economic development associated with increased private sector research and development; develoonment and funding of research to improve and expand Louisiana's agriculture and industry; coordination of state, federal, academic and business sector efforts to improve economic conditions within the state through economic development in the biotechnology sector. *http://www.labiotech.org/*

LumaCareTM Biotechnology

LumaCareTM is a scientific company developing instrumentation products in the field of biotechnology. The company has developed a specialized noncoherent light source for the field of photodynamic therapy (PDT) and for universal use with new Biotech pharmaceuticals (such as Photofrin). Photodynamic therapy is a safe, non-invasive treatment for numerous diseases, including psoriasis and other dermatological afflictions. *http://www.lumacare.com/*

Lynx Therapeutics, Inc.

Lynx applies its genomics, genetics, and proteomics technologies to accelerate the pace, broaden the scale, enhance the quality, and lessen the cost of the discovery and development efforts of customers in the pharmaceutical and agricultural biotechnology industries. Lynx has developed, and continues to develop, unique, proprietary processes aimed at handling and analyzing, simultaneously, very large numbers of DNA molecules or fragments in complex biological samples. At the core of these processes is Lynx's Megaclone™ technology. This is a micro-bead technology that allows both the simultaneous collection of millions of clones on as many micro-beads, and subsequent simultaneous manipulation or analyses of these clones. Applications include the physical extraction of genes (whether known or unknown) that are differentially expressed between samples (Megasort™ technology), the characterization of gene expression within a sample by Massively Parallel Signature Sequencing (MPSS®) technology, and a novel, highly efficient means for genotyping very large numbers of genetic markers or single nucleotide polymorphisms (SNPs), simultaneously, against large populations of genomes (Megatype™ technology). *http://www.lynxgen.com/*

Macromolecular Bioscience

The journal Macromolecular Bioscience covers research across the field of chemistry, physics, biotechnology, and biomedical applications of biomaterials including biopolymers (polymers from natural and renewable resources, such as proteins, peptides, carbohydrates, nucleic acids). Tables of contents from Jan 2001, no.1 are available free. Full-text articles are available with institutional or personal subscriptions. The site provides information on the editorial board, aims & scope, author guidelines, subscription details, and referee services. The intended audience is polymer scientists, chemists, physicists, and researchers "focused to the interface between chemistry, biology and medicine". The journal is published monthly by Wiley and is available online through Wiley InterScience.

Mendel Biotechnology

Mendel is a pioneer in the application of functional genomics to the study of plant genes. Mendel's business objective is to discover and characterize the function of important plant genes and to develop and commercialize high-margin products based on these discoveries. These products will benefit customers in agriculture, healthcare, specialty ingredient, and consumer product markets. *http://www.mendelbio.com/*

Michigan Biotechnology Association (MBA)

The MBA, a not-for-profit membership association, was launched to increase the understanding of the benefits of biotechnology and to support the growth of bio-based businesses in Michigan. The development of infrastructure to assist in the growth of biotechnology businesses is the primary goal of MBA. MBA also disseminates information of interest to it's members. Its members include companies, universities, research institutes and individuals who are, or would like to become, involved in biotechnology. *http://www.michbio.org/*

Microbiology Network

Microbiology Network aims to provide a communication resource for microbiologists and biotechnologists within academia and industry. It provides directories of relevant user groups and mailing lists, a directory of useful computer files of interest to microbiologists, a product directory, and relevant Web links. The Microbiology Network also hosts the PMFList, an email list devoted to microbiology in regulated industries (pharmaceutical, food and personal care). Sponsorship is provided by a number of commercial microbiology companies. Made available on the Web by Scott Sutton of Vectech Pharmaceutical Consultants.

Micrologix Biotech Inc.

Micrologix Biotech is a public biopharmaceutical company focusing on the discovery, development and commercialization of innovative drugs to treat life-threatening diseases. Our drug candidates are based on the development of improved derivatives of naturally occurring peptide compounds found in the defense systems of virtually all life forms. These powerful proprietary compounds are remarkably effective at killi.., the multidrug-resistant microbes responsible for the most clinically important infectious diseases. *http://www.mbiotech.com/*

National Biotechnology Information Facility

The National Biotechnology Information Facility (NBIF) provides a single-point access to a vast store of widely distributed biotechnology data including approximately 7,000 links to biotechnology resources. The NBIF encourages information sharing between researchers in Academia, Industry, Government, and the Department of Defense, promotes biotechnology in Historically Black Colleges and Universities and Minority Institutions (HBCU/MIs), and provides training in biotechnology. The NBIF is also active in developing new sources and types of biotechnology databases. http://www.nbif.org

National Center for Biotechnology Information (NCBI)

NCBI, a national resource for molecular biology information, creates public databases, conducts research in computational biology, develops software tools for analyzing genome data, and disseminates biomedical information, all for the better understanding of molecular processes affecting human health and disease. *http://www.ncbi.nlm.nih.gov/*

National Centre for Biotechnology Education

The National Centre for Biotechnology Education (NCBE) was founded in 1985, as part of the Department of Microbiology at the University of Reading. It aims to support the teaching of biotechnology in schools, providing training, resources and information for schools, industry, the public, and professional organisations. Information is provided here about events, materials, reviews of resources, protocols, safety information, technical advice, and links to other relevant resources. Made available on the Web by Reading University.

Nature Biotechnology

Nature Biotechnology, first published in 1996, is part of the Nature family of international journals. Nature Biotechnology publishes novel biological research with significant commercial potential in the pharmaceutical, medical, agricultural, and environmental sciences. Nature Biotechnology on the Web provides full-text services to subscribers of the monthly printed edition, while visitors and registered users (free registration on completion of a small form) have access to free contents and abstracts of the current issue, and to previous issues available in the archive section dating back to March 1996 (vol.14 issue 3). Nature Biotechnology is published by Macmillan Magazines Ltd.

NCBE

Safety in the School Laboratory: A collection of documents on safety, aimed at schools and colleges carrying out practical teaching in the field of biotechnology. This collection includes general safety guidelines, practical work with DNA, and safety guidelines relating to microbiology and biotechnology, handling solid and liquid enzymes, and plant and animal tissue culture. Published by the National Centre for Biotechnology Education, these guidelines apply within the UK only.

Neugenesis Corporation

Neugenesis Corporation is a biotechnology company focused on generating better biotherapeutics cost-effectively. To this end, Neugenesis has developed technology that has substantial value in discovering, improving, and producing complex proteins, especially heteromeric proteins. The company's work with the filamentous fungus *Neurospora crassa* has resulted in significant advances in two important areas of biotechnology research, recombinant protein production and combinatorial biology. *http://www.neugenesis.com/*

New Zealand Biotechnology Association

The Association was established to improve knowledge of biotechnology in New Zealand; to represent the interests of those concerned with biotechnology in New Zealand; to provide a forum for interaction among the disciplines involved in biotechnology; and to foster an environment conducive to the growth of biotechnology-based industries in New Zealand. *http://www.biotech.org.nz/*

North Carolina Biotechnology Center

The Center's mission is to provide long-term economic benefit to North Carolina through support of biotechnology research, development and commercialization statewide. It works toward that mission by assisting business, education, and science. The Center is active at all points in the movement of biotechnology from ideas to marketable products. It works with academic researchers, entrepreneurs and company scientists, educators, governmental policy-makers and all other parties involved in building a thriving biotechnology industry. *http://www.ncbiotech.org/*

Northern Bioidentification Service Ltd.

Northern Bioidentification Service Ltd. is a biotechnology company providing DNA analysis techniques to fish and wildlife conservation managers. The company works with animal conservation managers and researchers to protect and improve natural stocks of fish and wildlife. Through the use of high-quality DNA typing and population genetic analysis, the company aims to provide powerful and cost-effective tools used in conservation management. *http://www.mts.net/~northern/*

NRC Biotechnology Research Institute

BRI is located in Montréal, Canada, and maintains advanced facilities to carry out collaborative research projects in molecular biology and biochemical engineering, including: radiocrystallography equipment, an ion spray mass spectrometer, NMR spectrometers, computer-assisted molecular design, confocal light scanning microscopy system, analytical chemistry laboratory (for the environmental sector), a 1,500 m^2 bioprocess facility with fermenters ranging in size from 3.5 L to 1,500 L, and 350 m^2 of clean rooms for separation and purification (class 10,000 and 100,000). BRI concentrates its resources on collaborative projects with Canadian industry, particularly companies working in the pharmaceutical and resource sectors. *http://www.bri.nrc.ca/*

NRC Plant Biotechnology Institute

The mission of the Plant Biotechnology Institute is to perform, assist and promote strategic research in plant biotechnology, to improve and diversify Canadian industry, and to strengthen Canada's competitive position in the global economy by fostering an entrepreneurial culture. The Institute performs strategic research in key areas of plant biotechnology while maintaining a world-class research calibre; ensures industrial relevance of its research programs; nurtures innovation through efficient transfer of new biotechnologies and their products to commercialization; provides infrastructure and expertise; informs and transfers discoveries in plant biotechnology to all interested parties in Canada; and trains and develops highly qualified personnel in selected areas of plant biotechnology. *http://www.pbi.nrc.ca/*

Onyvax Limited

Onyvax is an early-stage biotechnology company focused on the development of cancer therapies that employ the selectivity and power of the immune system to seek and destroy tumor cells. *http://www.onyvax.com/*

Osmetech

Osmetech is dedicated to the rapid and inexpensive detection of micro-organisms such as infectious bacteria, fungi and moulds through the application of multi-sensor array technology. *http://www.aromascan.com/*

Pew Initiative on Food and Biotechnology

Established to be an independent and objective source of credible information on agricultural biotechnology for the public, media and policymakers, the Pew Initiative on Food and Biotechnology is funded through a grant from The Pew Charitable Trusts to the University of Richmond. The Initiative advocates neither for, nor against, agricultural biotechnology but instead is committed to providing information and encouraging debate and dialogue so that consumers and policymakers can make their own informed decisions. *http://pewagbiotech.org/*

Phytera, Inc.

Phytera, Inc. is an international biotechnology company that applies the genetic and chemical diversity of natural species to the discovery of new pharmaceutical, agricultural and other products. Phytera has developed proprietary cell culture technologies for plant species (ExPAND®) and marine microbes (μMARINE®) that provide the Company with unique and greatly expanded access to the genetic and chemical diversity of each species. To date, the Company has assembled the largest library of plant cells in culture in the world and also has an extensive library of marine microbial cultures. The ExPAND® and μMARINE® technologies involve extensive manipulation of the cell cultures to simulate many different environments. This combinatorial biology approach turns on silent genes in the cultured cells and results in the expression of greatly expanded chemical diversity. Phytera captures the genes and expanded chemical diversity in its proprietary databases and libraries. *http://www.phytera.com/*

Politics of GM Food

Risk, Science and Public Trust: A report published by the Global Change Development Programme. This report addresses the issue of genetically modified food, examining the politics, ethics and scientific thinking surrounding its implementation in the UK. The emphasis of the report is focused on the UK government's handling of the GM food debate. References are provided and many are available via the Web. The report can be downloaded free in either PDF (requiring Adobe Acrobat Reader software) or HTML. The report is aimed at anyone with an interest in the UK GM food debate. This was a ten year research programme established in 1991, and was funded by the Economic and Social Research Council.

Princeton Separations, Inc. (PSI)

PSI is a privately-owned, manufacturing-based biotechnology company specializing in molecular separations. PSI manufactures various successful products including: CENTRI·SEPTM spin columns, used in DNA sequencing; PANAGEL® high-resolution electrophoresis systems, used in analysis of proteins in human body fluids; CENTRI·SPINTM columns, used in the purification of a variety of biomolecules (DNA, proteins and carbohydrates) and PSI CLONE Plasmid Prep kit for purification of plasmid DNA. *http://www.prinsep.com/*

Progen Industries Limited

Progen Industries Limited is an Australian biotechnology company specialising in the manufacture and supply of molecular biology reagents and kits. Involved in pharmaceutical research and development, Progen recently commenced its second Phase I clinical trial with its lead compound (PI-88) for treatment of solid tumors. Progen is also involved in the development of a number of other potential biopharmaceuticals. *http://progen.com.au/*

Q-One Biotech

Q-One Biotech is a world-renowned authority in providing specialist safety testing and process validation services, cGMP contract manufacturing of cell banks, vaccines and gene therapy products to the biopharmaceutical industry. Q-One operates in full compliance with GMP and GLP standards. *http://www.q-one.com/*

Quebec Bio-Industries Association

The Quebec Bio-Industries Association, which counts close to 200 innovative companies as members, is committed to supporting the growth and vitality of Quebec's biotechnology companies. The QBA's member companies are active in research and manufacturing of therapeutic and diagnostic products, vaccines and biological products of human and animal health; in research and primary production in the areas of agriculture, aquaculture, secondary processing and beverages; and in research and production in the waste treatment, biomass processing, environmental rehabilitation, mining and peat bog sectors. *http://www.aqb.qc.ca/*

Quebec Biotechnology Innovation Centre (QBIC)

QBIC is the first business incubator specializing in biotechnology to emerge in Québec. This non-profit organization offers researcher-entrepreneurs in Québec and elsewhere, an environment conducive to their research and development activities, while promoting the acquisition and transfer of management skills. The goal of the QBIC is to promote the development of young innovative companies working in the field of biotechnology applied to the health, environment, agri-food and forestry sectors. To do this, the QBIC makes specialized equipment infrastructures available to these companies and offers them business services and counseling in line with their needs. By supporting the start-up of companies in this new economic sector, the QBIC and its partners are contributing to job creation in Québec and the emergence of a generation of entrepreneurs in the biotechnology sector. *http://www.cqib.org/indexe.htm*

Recombinant Capital (ReCap)

The purpose of the ReCap website is to provide a resource to biotechnology executives who are involved with financing, business development, licensing and other strategic planning essential to the development and commercialization of biotechnology, pharmaceuticals, diagnostics and medical devices. *http://www.recap.com/*

Reliable Biopharmaceutical Corporation

Reliable Biopharmaceutical Corporation is a manufacturer of active pharmaceutical ingredients, nucleic acids, biochemicals, and enzymes for the pharmaceutical and biotechnology industries. The company isolates deoxynucleosides and deoxynucleotides from Salmon DNA. The company's nucleotides and nucleosides are of the highest purity available. They also synthesize DMT protected 2'O methyl ribonucleosides and deoxynucleosides. After these protected deoxy and ribo nucleosides are converted to phosphoramidite derivatives, they can be incorporated into the new antisense therapeutic and diagnostic compounds.

Reticula Bio

Reticula Bio is a research and development consortium that specializes in advanced technology and biopharmaceutical research and development. The company's two divisions reflect its broad research interests. The biosciences division investigates and develops novel pharmaceutical, biotechnology and agricultural products and the advanced technologies division specializes in immersive technology, robotics and computing applications. *http://www.reticulabio.com/*

Scientific Advisory Committee on Genetic Modification

Home page for the Scientific Advisory Committee on Genetic Modification, an advisory committee of the Health & Safety Commission. The role of the SACGM is to "advise the Health and Safety Commission and Executive and the Secretary of State for the Environment, and other Ministers and bodies, as appropriate, on all aspects of the human and environmental safety of the contained use of genetically modified organisms." Full-text of agendas and newsletters are provided here, as well as information on the contained use of genetically modified organisms. Made available on the Web by the UK government's Health and Safety Commission.

Seikagaku Corporation

Seikagaku Corporation has evolved into a major contributor in the biotech industry specializing in pharmaceutical products such as ARTZ®, research biological and diagnostics such as Endospecy®, as well as pursuing R&D efforts in developing technologies based on Glycoconjugates. *http://www.seikagaku.com/*

Senetek PLC

Senetek PLC is a science-driven biotechnology company

that develops, manufactures and markets proprietary solutions for aging-related healthcare problems. Senetek business divisions include three separate, distinct and significant opportunities in biopharmaceuticals, autoinjectable drug delivery systems, and cosmeceuticals. *http://www.senetekplc.com/*

Society for Industrial Microbiology

The Society for Industrial Microbiology (SIM) is a professional society involved in the development of microbiology with the particular application to industrial products, biotechnology, materials and processes. SIM liases between the various fields of theoretical and applied microbiology through workshops, meetings and publications. The main subject areas includes fermentation, bioremediation/biodeterioration, recombinant DNA technology, secondary metabolism, and biotransformation. The site provides information on careers in microbiology, meetings, membership, as well as publications and resources in PDF (requiring Adobe Acrobat Reader).

Sunnyday Environmental & Bio-Tech Services Ltd.

While Sunnyday has the staff, expertise and experience to handle a wide variety of projects, their specialty is the design and construction of complete utility systems for pharmaceutical research labs and cGMP pilot plant facilities which demands specific knowledge with regard to cGMP and FDA requirements. The ultimate goal is not only high performance, but also validation of utility systems. All the necessary equipment, instrumentation and control, material and labor are supplied to meet the needs of the company's projects. For pure steam/WFI systems the company performs the orbital stainless steel welding using computer controlled automatic equipment to secure consistent quality. *http://www.sunnyday-biotech.com/index.html*

Techniclone Corporation

Techniclone Corporation is a biotechnology company focused on the research, development and commercialization of novel cancer therapeutics in two principal areas: 1) direct tumor targeting agents for the treatment of refractory malignant lymphomas and 2) collateral indirect targeting agents for the treatment of solid tumors. *http://www.techniclone.com/*

Tissue Engineering Society International

The home page of the Tissue Engineering Society International (TESI). The society promotes education and research in the field of tissue engineering through meetings, publications and collaboration. The Web site provides access to the society's annual newsletter, details of relevant publications and information on forthcoming meetings. Subscribed members of the society can also access TESIExchange, a network designed to advance the field of tissue engineering through increased collaboration and knowledge sharing.

Toronto Biotechnology Initiative (TBI)

The TBI is a non-profit organization committed to promoting the growth of biotechnology in Toronto and the surrounding region. While biopharmaceuticals and diagnostics are core strengths of this community, it also has broad research capability and growing commercial activity in agri-food biotechnology as well as the emerging resource and environmental biotech industries.

Transgenomic, Inc.

Transgenomic, Inc. is a global company with a core biotechnology business that includes principal technologies for DNA/RNA and analysis and purification. This diverse technology base enables world-class research with technology and instrumentation that allows scientists to get the answers they seek faster and with greater confidence. Principal application areas include: genetic mutation screening and detection, plus new paths in microbial, pharmaceutical, biotechnology, forensic, plant and animal research.

Trends in biotechnology

Trends in Biotechnology is a journal published monthly by Elsevier Science, and hosted online by BioMedNet. The journal aims to keep its readers up-to-date and well informed about all areas of biotechnology, from molecular genetics to biochemical engineering. The journal's principal content is reviews, but it has other sections that cover workshops, meeting reports, novel techniques, ethical issues, and business issues relevant to all biotechnologists. Tables of contents are available free, with abstracts and full-text articles available (from 1998, in PDF) by per-article fee, personal subscription or to licensed institutions. Membership of BioMedNet is free and registration takes a few minutes. The journal is aimed at all biotechnologists who take an "integrated, multidisciplinary approach to the applied biosciences."

University of Minnesota Biocatalysis/biodegradation Database

Home page of the University of Minnesota Biocatalysis/ Biodegradation Database of microbial biocatalytic reactions and biodegradation pathways primarily for xenobiotic, chemical compounds. The goal of the UM-BBD is to provide information on microbial enzyme-catalyzed reactions that are important for biotechnology, including biocatalysis leading to specialty chemical manufacture, and biodegradation of environmental pollutants. The site provides browsable lists of pathways, reactions, compounds, and microorganisms. There is also a link to Biocatalysis and Biodegradation, a graduate-level course offered over the Internet. The database has been supported by the University of Minnesota Biological Process Technology Institute, the National Institutes of Health, and National Science Foundation.

Vical Incorporated

Vical develops biopharmaceutical products based on patented naked DNA gene transfer technologies for the prevention and treatment of life-threatening diseases. The company currently focuses its development on innovative cancer therapies designed to induce an immune response against cancer cells without causing serious side effects.

Virginia BioTechnology Research Park

The Virginia BioTechnology Research Park is a technology center dedicated to fostering development of Virginia's biosciences industry through technology transfer, new business formation, expansion of existing businesses, and business attraction.

Bibliography

Collins, F. & Galas,D. A new five-year plan for the US Human Genome Project. *Science* 262, 43–46 (1993).

Drews, J.& Ryser, S. The role of innovation in drug development.

Fuller, B. P. *et al.* Policy forum: Ethics — privacy in genetics research. *Science* 285, 1359–1361 (1999).

Gabriel, S. B. *et al.* The structure of haplotype blocks in the human genome. *Science* 296, 2225–2229 (2002).

Gardner, M. J. *et al.* Genome sequence of the human malaria parasite *Plasmodium falciparum*. *Nature* 419, 498–511 (2002).

Guttmacher and Mark S.Guyer are at the National Human Genome Research Institute, National Institutes of Health, Bethesda, Maryland 20892, USA.

Holt, R. A. *et al.* The genome sequence of the malaria mosquito *Anopheles gambiae*. *Science* 298, 129–149 (2002).

Liechty KW, MacKenzie TC, Shaaban AF, et al. Human mesenchymal stem cells engraft and demonstrate site-specific differentiation after in utero transplantation in sheep. Nat Med 2000, 6:1282-6.

Lin Y, Weisdorf DJ, Solovey A, Hebbel RP. Origins of circulating endothelial cells and endothelial outgrowth from blood. J Clin Invest 2000;105:71-77.

Lynch, H. T. & de la Chapelle, A. Genomic medicine:

hereditary colorectal cancer.*N.Engl. J.Med.* 348, 919–932 (2003).

Lysaght, M.J., and Hazlehurst, A.L., "Private Sector Development of Stem Cell Technology and Therapeutic Cloning," *Tissue Engineering* 9(3): 555-561 (2003).

McCarthy PM, Rose EA, Macmanus Q, et al. Clinical experience with the Novacor ventricular assist system. J Thorac Cardiovasc Surg. 1991; 102:578-587.

McEwan PE, Gray GA, Sherry L, Webb DJ, Kenyon CJ. Differential effects of angiotensin II on cardiac cell proliferation and intramyocardial perivascular fibrosis in vivo. Circulation 1998;98:2765-2773.

Mendel, G.Versuche über Pflanzen-Hybriden.*Verhandlungen des naturforschenden Vereines, Abhandlungen, Brünn* 4, 3–47 (1866).

Milki, A.A. et al., "Incidence of monozygotic twinning with blastocyst transfer compared to cleavage-stage transfer" *Fertility and Sterility* 79(3): 503-506 (2003).

National Institutes of Health Revitalization Act of 1993, Pub. L. No. 103-43, § 121(c), 107 Stat. 122 (1993), repealing 45 C.F.R. § 46.204(d).

Oz MC, Argenziano M, Catanese KA, et al. Bridge experience with long-term implantable left ventricular assist devices. Are they an alternative to transplantation? Circulation. 1997; 95: 1844-1852.

Park, T.S., Hong, Y.H., Kwon, S.C., Lim, J.M., and Han, J.Y. 2003. Birth of germline chimeras by transfer of chicken embryonic germ (EG) cells into recipient embryos.

Pittenger MF, Mackay AM, Beck SC, et al. Multilineage potential of adult human mesenchymal stem cells. Science 1999, 284: 143-147.

Rothenberg, K. *et al.* Genetic information and the workplace: Legislative approaches and policy challenges. *Science* 275, 1755–1757 (1997).

Smythe GM, Grounds MD (2000) Exposure to tissue culture conditions can adversely affect myoblast behaviour in vivo in whole muscle grafts: implications for myoblast transfer therapy. Cell Transplant 9: 379-393.

Stockwell, B.R. Chemical genetics: Ligand-based discovery of gene function. *Nature Rev. Genet.* 1, 116–125 (2000).

Index

C

D

E

F

G

H

I